GMAW/FCAW Handbook

First Edition

William H. Minnick
James Mosman

Welding Technology Department Chair
Odessa College, Texas

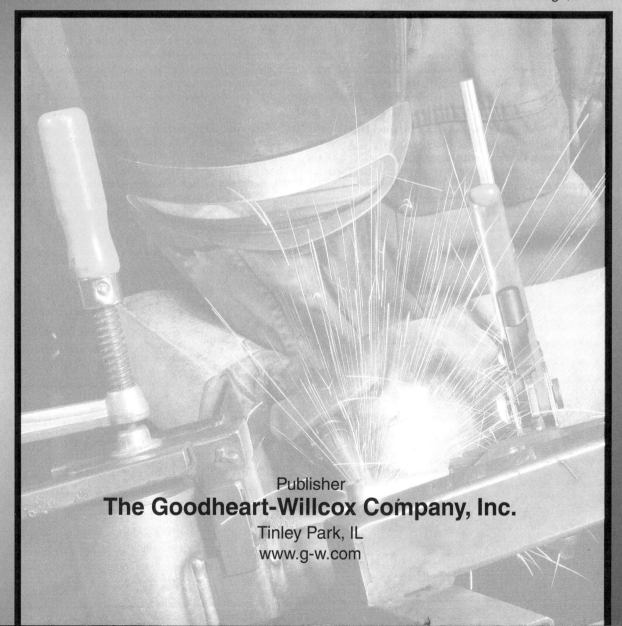

Publisher
The Goodheart-Willcox Company, Inc.
Tinley Park, IL
www.g-w.com

Library of Congress Catalog Card Number 2016028009

ISBN 978-1-63126-365-1

1 2 3 4 5 6 7 8 9 – 18 – 21 20 19 18 17 16

Cover image: Aumm graphixphoto/Shutterstock.com
Chapter opener image: B Brown/Shutterstock.com

Library of Congress Cataloging-in-Publication Data

Names: Minnick, William H., author. | Mosman, James, author.

Title: GMAW/FCAW handbook / William H. Minnick, James Mosman, Welding

 Technology Department Chair, Odessa College, Texas.

Other titles: Gas metal arc welding/flux-cored arc welding handbook

Description: 1st edition. | Tinley Park, Illinois: The Goodheart-Willcox

 Company, Inc., [2018] | Includes bibliographical references and index.

Identifiers: LCCN 2016028009 | ISBN 9781631263651

Subjects: LCSH: Gas metal arc welding--Handbooks manuals, etc. | Flux

 (Metallurgy)--Handbooks manuals, etc.

Classification: LCC TK4660 .M535 2018 | DDC 671.5/212--dc23 LC record available at
 https://lccn.loc.gov/2016028009

Introduction

This edition of the *GMAW/FCAW Handbook* is an updated and compiled version of two textbooks by William H. Minnick previously published by The Goodheart-Willcox Company, Inc. This new edition provides a simple but complete introduction to the various welding processes of GMAW and FCAW. The handbook covers the principles, equipment, techniques, processes, and safety involved with these welding methods.

Several portions of the previous editions have been completely rewritten and updated according to current industry standards. This edition also includes all new color photos, the vast majority of which were taken by the author. In addition, illustrations have been redesigned, adding color and detail for improved clarity and learning value.

Gas metal arc and flux cored arc welding processes are widely used in manufacturing and construction industries. These processes continue to evolve and be refined as new technology is developed. Careful study of this textbook and practice with the techniques presented allows welding students to progress from novices to proficient entry-level welders. Advanced students and professional welders will find this text an informative and valuable resource throughout their welding career. Instructors can rely on this textbook to provide accurate, up-to-date, and thorough instruction that is designed to prepare every student for a welding career.

Students: As you continue to advance with your welding skills, remember that every weld you make is your personal signature or trademark. Take pride in your skills and make each weld as if your career depended on it!

James Mosman

Reviewers

The author and publisher wish to thank the following industry and teaching professionals for their valuable input into the development of *GMAW/FCAW Handbook*.

Tim Baber
College of the Canyons
Santa Clarita, California

Brian Bennett
Hill College
Hillsboro, Texas

Dustin Canady
University of Arkansas Community College
Morrilton, Arkansas

Brian Yarrison
York County School of Technology
York, Pennsylvania

Acknowledgments

The publisher would like to thank the author for the numerous high-quality and technically accurate photographic images that he contributed to this new text.

The author and publisher also thank the following companies and organizations for their contribution of resource material, images, or other support in the development of *GMAW/FCAW Handbook*.

American Society of Mechanical Engineers

American Welding Society

David W. Andrews

B & J Welding Supply Ltd.– Midland, TX

Steve Goff

Justin Goodson

J & W Services Inc.

Lawson Welding & Fabrication LLC

Lincoln Electric Co.

Matheson Welding Supply – Odessa, TX

McAfee Machine Inc.

Miller Electric Mfg. Co.

Odessa College

United Rentals Inc.

Westair Gas and Equipment L.P. – Odessa, TX

G-W Integrated Learning Solution

Together, We Build Careers

At Goodheart-Willcox, we take our mission seriously. Since 1921, G-W has been serving the career and technical education (CTE) community. Our employee-owners are driven to deliver exceptional learning solutions to CTE students to help prepare them for careers. Our authors and subject matter experts have years of experience in the classroom and industry. We combine their wisdom with our expertise to create content and tools to help students achieve success. Our products start with theory and applied content based upon a strong foundation of accepted standards and curriculum. To that base, we add student-focused learning features and tools designed to help students make connections between knowledge and skills. G-W recognizes the crucial role instructors play in preparing students for careers. We support educators' efforts by providing time-saving tools that help them plan, present, assess, and engage students with traditional and digital activities and assets.

Student-Focused Curated Content

Goodheart-Willcox believes that student-focused content should be built from standards and/or accepted curriculum coverage. AWS SENSE QC10 standards are addressed in this text. *GMAW/FCAW Handbook* also uses a building block approach with attention devoted to a logical teaching progression that helps students build upon their learning. We call on industry experts and teachers from across the country to review and comment on our content, presentation, and pedagogy. Finally, in our refinement of curated content, our editors are immersed in content checking, securing, and sometimes creating figures that convey key information, and revising language and pedagogy.

Features of the Textbook

Features are student-focused learning tools designed to help you get the most out of your studies. This visual guide highlights the features designed for the textbook.

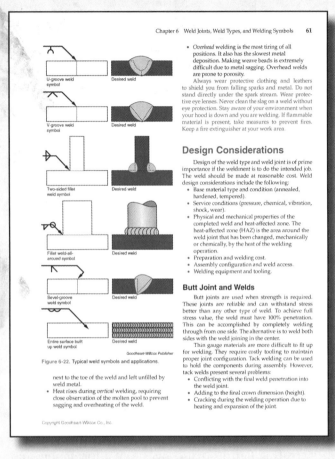

Safety information is printed in a different color font to alert you to potentially dangerous materials and practices.

Objectives clearly identify the knowledge and skills to be obtained when the chapter is completed.

Technical Terms list the key terms to be learned in the chapter. The definition of each term is listed in the Glossary in the back of the book to help you master welding vocabulary.

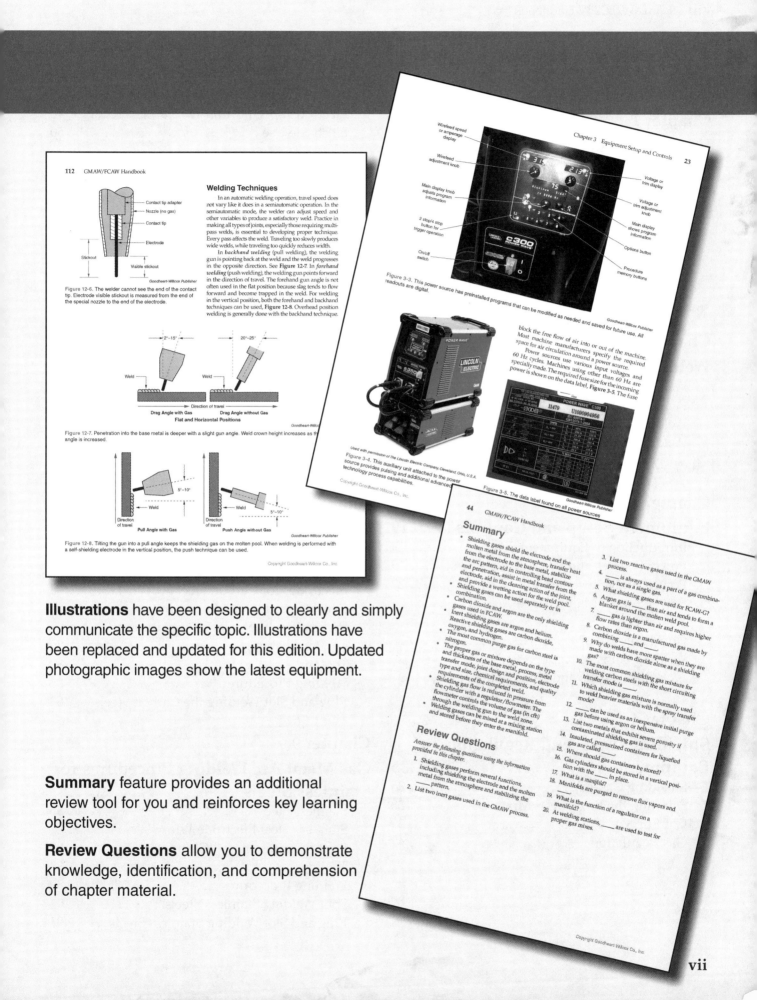

Illustrations have been designed to clearly and simply communicate the specific topic. Illustrations have been replaced and updated for this edition. Updated photographic images show the latest equipment.

Summary feature provides an additional review tool for you and reinforces key learning objectives.

Review Questions allow you to demonstrate knowledge, identification, and comprehension of chapter material.

Contents

About the Authors

James Mosman was introduced to welding as part of the Agriculture program while attending high school in his hometown of Alden, New York. After a four-year tour with the US Army, Mr. Mosman attended and graduated at the top of his class from El Paso Trade School as a certified welder. He moved to Odessa, Texas to work in the oil and gas manufacturing industry in 1981. Mr. Mosman spent the next 20 years as a welder, working both in the field and in manufacturing facilities.

Mr. Mosman earned his AAS degree from Odessa College and a BA degree from University of Texas Permian Basin. He is also a certified welding inspector through the American Welding Society and an NCCER Certified Master Trainer.

Mr. Mosman is an associate professor of welding technology and department chair of Industrial Technology at Odessa College, where he has taught since 1998. Under his direction, the Odessa College welding program has grown from 45 students to over 200 enrolled each semester. He served as Program Coordinator of a Department of Labor four-year grant, which provided introductory welding courses to over 700 students. He also continues to work as a CWI, welding consultant, and trainer for several companies throughout Texas.

Mr. Mosman is a member of the *Practical Welding Today* Editorial Advisory Committee and has contributed to several articles for this magazine. He also serves on the City of Odessa Public Art Committee and has built metal sculptures in his artistic moments. In his spare time, he enjoys traveling throughout the United States, Europe, and the West Texas back roads on his Harley-Davidson.

William H. Minnick, realizing the need for a specialized type of welding text for instructors and students, drew on his many years of experience as a welder, welding engineer, and community college instructor to develop welding texts for training future welders. The author's career in industry included experience welding on jet engines, missiles, pressure vessels, and nuclear reactors. He developed the welding procedures for, and welded, the first titanium pressure vessel for the Atlas missile program. He has written many technical papers, including articles on the research and development of procedures for welding exotic materials and on the modification of existing welding processes for automatic and robotic applications.

Mr. Minnick taught all phases of welding and metallurgy in community colleges for more than 20 years. He developed welding certificate and degree programs. In addition to this textbook, Mr. Minnick has authored the *Gas Tungsten Arc Welding Handbook*.

Chapter 1

Introduction to Gas Metal Arc Welding and Flux Cored Arc Welding

Objectives

After studying this chapter, you will be able to:
- ❏ Define gas metal arc welding (GMAW).
- ❏ Define flux cored arc welding (FCAW).
- ❏ Describe the four GMAW deposition modes.
- ❏ Describe the two types of FCAW processes.

Technical Terms

automatic welding
constant current (CC)
constant voltage (CV)
electrode wire
flux cored arc welding (FCAW)
gas metal arc welding (GMAW)
globular transfer
power source
pulsed spray transfer
semiautomatic welding
shielding gas
short circuiting transfer
spray transfer
welding gun
wire feeder

GMAW and FCAW

Two welding processes commonly used in today's manufacturing industry are gas metal arc welding (GMAW) and flux cored arc welding (FCAW). Both processes use a power source/welding machine with a constant voltage (CV) output and a wire feeding mechanism to deliver a continuous wire electrode. These processes are also referred to as *wire welding, MIG/MAG welding,* and *flux-core welding.* See **Figure 1-1.**

History

The idea of using inert (not chemically reactive) gas to shield molten metal was introduced in the 1920s. However, it was not until 1948 that a patent was issued

Goodheart-Willcox Publisher

Figure 1-1. A welder prepares to weld a test plate with the GMAW process.

for a consumable electrode welding machine and the metal inert gas (MIG) welding process began. It then took another 10 years before the concept became commercially available to industry.

Initially, the inert gases helium and argon were the primary shielding options for the MIG welding process. The introduction of carbon dioxide as a single shielding source and its use in a mixture with inert gases led to the term *gas metal arc welding*, although the process is still often referred to as MIG welding. During early use of the process, several companies developed their own trade names. Miller Electric had Millermatic, Lincoln Electric had Squirt Welder, and Air Reduction Company had Aircomatic machines.

Arthur Bernard is attributed with the development of the flux cored arc welding process in 1954. With the introduction of CO_2 as a viable shielding gas in the mid 1950s, the use of a flux-filled tubular wire electrode was patented by National Cylinder Gas Company in 1957. This created a weld pool dual-shielded by the flux along with the gas shield. Further developments by Lincoln Electric Motor Company eliminated the need for the gas, and self-shielding FCAW electrode wire became available in 1959.

Both GMAW and FCAW developed around the concept of a continuously fed electrode wire, a constant voltage power source, and a shielding gas around the molten metal weld pool. These two processes continue to be further developed as electronic and robotic technology advances.

Definitions

As defined by the American Welding Society (AWS), *gas metal arc welding (GMAW)* is an electric arc welding process used to fuse metallic parts by heating them with an arc between a continuously fed welding wire electrode and the workpiece. See **Figure 1-2**. The electrode wire is small in diameter and is consumed during the process. It is melted as a result of the electric arc and becomes part of the weld. An externally supplied *shielding gas* protects the molten weld metal from contamination by the atmosphere.

Flux cored arc welding (FCAW) is similarly defined by the AWS as an electric arc welding process used to fuse metallic parts by heating them with an arc between a continuously fed welding wire electrode and the workpiece. The electrode used in this process is a tubular flux-filled wire. As the electrode melts, this flux creates a slag deposit on top of the weld bead to shield it from the atmosphere. This may be supplemented with an externally supplied shielding gas to help protect the molten metal from atmospheric contamination. See **Figure 1-3**.

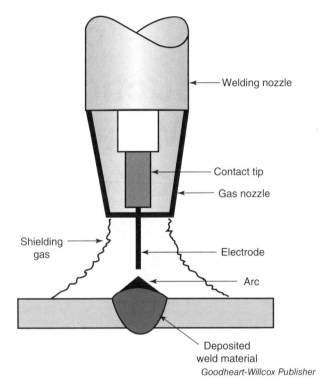

Goodheart-Willcox Publisher

Figure 1-2. In the GMAW process, an electrode wire creates an arc that is covered by a shielding gas.

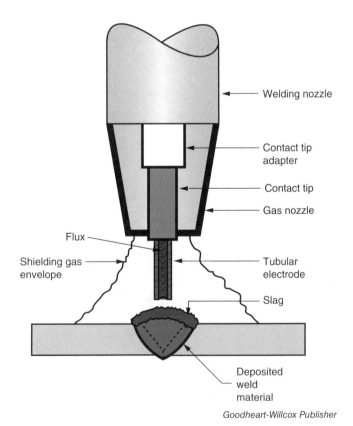

Goodheart-Willcox Publisher

Figure 1-3. FCAW uses a tubular electrode wire filled with a flux. This process may or may not use a shielding gas.

Both the GMAW and FCAW processes use a power source with a *constant voltage (CV)* system, sometimes referred to as a *constant potential* welding machine. A CV power source yields a large welding current change from a small arc voltage change. This power source is different from the **constant current (CC)** system used for shielded metal arc welding (SMAW) and gas tungsten arc welding (GTAW) processes. In CC power sources, a small current change occurs from a large arc voltage change. Multiprocess (CC-CV) power sources, which can switch between both types of systems, are available to weld all four of these processes. See **Figure 1-4**. The electrical current of the welding machine is set for direct current electrode positive (DCEP), also known as direct current reverse polarity (DCRP).

Goodheart-Willcox Publisher

Figure 1-4. A multiprocess CC/CV welding outfit capable of GMAW, FCAW, SMAW, and GTAW.

Equipment and Modes of Operation

The GMAW and FCAW processes are often used in semiautomatic welding and are also used in automatic welding. In *semiautomatic welding*, the weld is produced and controlled by a welder using a hand-held gun, and the *electrode wire* is fed from a wire feeder system. The wire is melted as a result of the electric arc and becomes part of the weld. In *automatic welding*, the welder operator uses specialized mechanical or computer-controlled equipment or robotics to complete the required weld. These wire-fed processes are ideal for automated welding.

The basic equipment used for GMAW and FCAW processes for either semiautomatic or automatic welding consists of the following:
- Power source with a constant voltage output.
- Wire feeder.
- Electrode wire.
- Shielding gas.
- Welding gun.
- Work clamp and cables.

A welding *power source* is the machine that supplies the current and voltage suitable for the welding process. The power source may be an electrical transformer based machine or an engine-driven generated electrical source. The *wire feeder* controls the feed speed rate of the electrode wire, the shielding gas solenoid, and the electrical current from the work to the power source. The *welding gun* directs the electrode wire and shielding gas to the weld and contains the trigger switch that controls the wire feeder operations. The work clamp and cable supply the electrical current from the power source to the part to be welded.

A welding outfit containing all of these components with a wire feeder separate from the power source is shown in **Figure 1-5**. Some welding outfits contain the wire feeder as part of the machine, **Figure 1-6**. Later chapters provide detailed explanations of welding system components.

GMAW Deposition Modes

The GMAW process has been categorized into four types (modes) of deposition relative to the welding voltage and metal transfer across the welding arc. These modes are distinguished by how the arc is created and how the electrode wire melts off and is deposited into the weld zone.

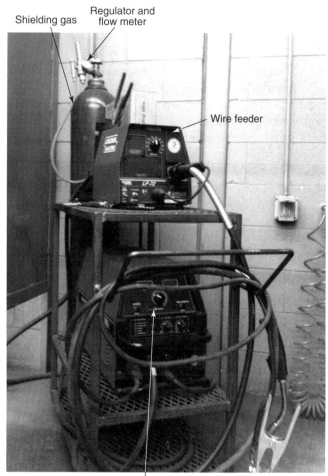

Figure 1-5. A complete welding outfit showing all of the components. In this system, the wire feeder is separate from the power supply.

Short Circuiting Transfer

In *short circuiting transfer,* the electrode wire short-circuits when it is fed into and contacts the workpiece. The short circuit causes the electrode to melt off and be deposited as molten metal into the weld joint. See **Figure 1-7**. This deposition mode uses relatively low voltages and amperages.

This process has the following advantages:
- Low heat input.
- Welds in all positions.
- Welds thin gauge materials.
- Easily controlled weld pool.
- Can be used to bridge gaps and fill holes as needed.

This process has the following disadvantages:
- Cannot be used where wind draft exceeds five miles per hour. (This is true for all GMAW deposition modes.)
- Uses small diameter wire that reduces metal deposition rates.
- Results in spatter (small globules of molten metal) when used with various gases and techniques.
- Cold starts of the weld and cold lap during welding that can occur due to the low heat input.
- Lack of penetration on thicker materials without proper technique or preparation.

Globular Transfer

In the *globular transfer* mode, the electrode burns off above or in contact with the workpiece in

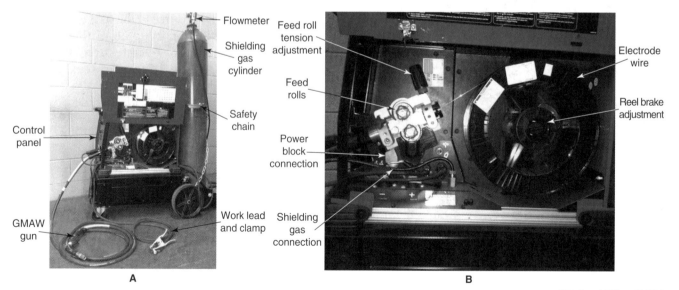

Goodheart-Willcox Publisher

Figure 1-6. This complete welding outfit has the wire feeder as part of the machine. It requires only a shielding gas cylinder and electrode wire. A—Welding outfit. B—Wire feeder.

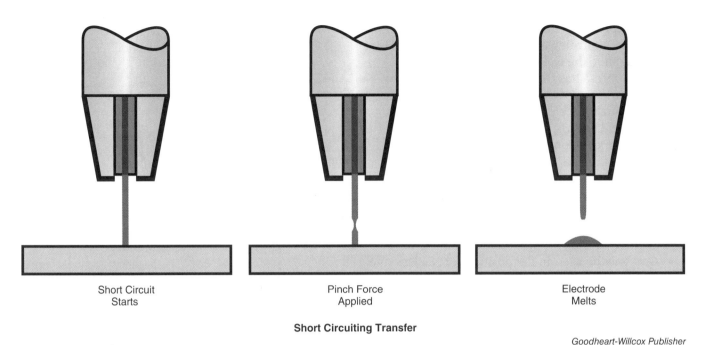

Short Circuit
Starts

Pinch Force
Applied

Electrode
Melts

Short Circuiting Transfer

Goodheart-Willcox Publisher

Figure 1-7. In the short circuiting mode of metal deposition, the electrode touches the workpiece and melts from 20 to 200 times per second.

an erratic globular pattern. See **Figure 1-8**. This mode results from the use of a voltage range between short circuiting and spray transfer ranges. Globular deposition is generally undesirable and not widely used in today's industry.

This process has the following advantages:
- Higher deposition rate.
- High-quality welds with proper procedures.

This process has the following disadvantages:
- Considerable smoke and spatter.
- Rough weld appearance.
- Can only be used in the flat position.

Spray Transfer

The *spray transfer* mode of deposition uses a higher voltage range than is used for short circuiting and

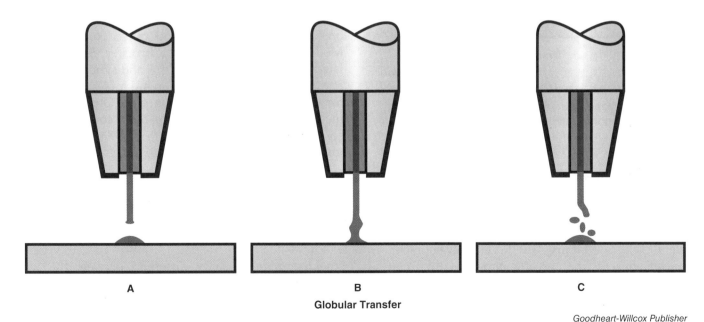

A

B

C

Globular Transfer

Goodheart-Willcox Publisher

Figure 1-8. The globular mode of deposition. A—An electrode drop forms and detaches due to high temperature of the arc. B—The electrode contacts the workpiece before melting. C—The electrode drop may disintegrate and cause spatter.

globular deposition. The electrode is melted off above the workpiece, forming a fine spray of molten metal that is deposited into the weld zone. See **Figure 1-9**.

This process has the following advantages:
- Very little or no spatter.
- High-quality weld metal deposition.
- Fast weld speeds.
- Deep penetration.
- Excellent results with automatic or robotic applications.
  This process has the following disadvantages:
- Large and very fluid weld pool.
- Higher heat input.
- Limited to use on thicker materials.
- Can only be used in flat position for steel groove welds.
- Can only be used on flat or horizontal steel fillet welds.

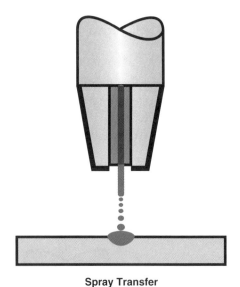

Spray Transfer

Goodheart-Willcox Publisher

Figure 1-9. In the spray transfer mode, the electrode is fed into the weld zone, melted off above the workpiece, and deposited randomly into the weld pool.

Pulsed Spray Transfer

In the *pulsed spray transfer* deposition mode, a drop of molten filler metal is melted off or pulsed from the electrode wire at a controlled rate and time in the weld cycle. See **Figure 1-10**. Recent developments in power sources have greatly improved this method of GMAW deposition and broadened its possible applications.

This process has the following advantages:
- Greater metal deposition with lower heat input.

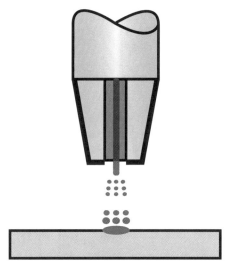

Pulsed Spray Transfer

Goodheart-Willcox Publisher

Figure 1-10. In the pulsed spray transfer mode, drops of filler material are pulsed from the end of the electrode at a controlled time in the welding cycle.

- Lower amperages reduce warpage and metal distortion.
- Virtually no spatter.
- May be used on most weldable metals.
  This process has the following disadvantage:
- Need for special power supplies, wire-feeder systems, or special pulsing units that attach to existing GMAW power sources.

Other Deposition Modes

Recently developed deposition modes are based on both the short circuit and pulsed technologies. Modes such as Surface Tension Transfer®(STT®), Regulated Metal Deposition (RMD™), and synergic pulsed and adaptive pulsed deposition are increasingly used for welding open root gaps on pipe. Other developments, such as pulsed AC capability for aluminum, continue to be integrated into the latest welding machines as technology and research evolve.

FCAW Process Types

Two processes can be used for flux cored arc welding—the self-shielded process and the gas-shielded arc process. The difference between the two processes is the way the weld pool is shielded from the atmosphere. Refer to **Figure 1-3**.

Self-Shielded (Open Arc) Process

In the self-shielded flux cored arc welding (FCAW-S) process, also known as the *open arc* process, all of the fluxing ingredients required for the shielding of the molten weld metal are included in the core material of the electrode. The flux creates a shielding gas coverage when heated and a protective slag coating when it cools to help protect the weld pool from atmospheric contamination and to slow its cooling rate.

The advantages of this process are as follows:

- Ability to use the process outdoors, even in slightly windy conditions.
- Eliminates the need for costly shielding gas.

The disadvantages of this process are as follows:

- Produces slag, which must be removed between passes.
- Produces more smoke than GMAW.
- Not as portable as SMAW.
- Requires more equipment than SMAW.

Gas-Shielded (Dual-Shield) Process

The gas-shielded flux cored arc welding (FCAW-G) process, also known as the *dual-shield* process, uses carbon dioxide alone or in combination with argon as a mixture to shield the weld pool and arc stream from the outside atmosphere. The shielding gas provides additional protection from atmospheric contamination to that provided by the flux in the electrode.

The advantages of this process are as follows:

- Very good metal deposition rates.
- Very good metallurgical values.
- Electrode flux cannot be broken off as occurs in shielded arc welding (SMAW).
- Good visibility of the molten weld pool.
- Welds can be made in all positions.

The disadvantages of this process are as follows:

- Produces slag, which must be removed between passes.
- Produces more smoke than GMAW.
- Not as portable as SMAW.
- Cannot be used outdoors if wind exceeds 5 mph.
- Requires more equipment than SMAW.

Which Process to Use?

The similarities and differences and advantages/disadvantages between GMAW and FCAW can make choosing a process confusing. To determine the best process for a particular application, it is necessary to consider some of the variables. These include material type and thickness, shielding gas availability, wire type and size, voltage rating of the power source, location of the work, and required weld appearance. All of these variables affect the decision to use a solid wire or flux cored wire process. Information provided in later chapters can help a welding operator make the best decision.

Summary

- Gas metal arc welding (GMAW) is an arc welding process using an arc between a continuous solid filler metal electrode and the weld pool. The weld pool is protected from atmospheric contamination by an externally supplied shielding gas.
- Flux cored arc welding (FCAW) is an arc welding process using an arc between a tubular flux-filled wire electrode and the weld pool. The weld pool is protected by a shielding gas created from a flux within the tubular electrode. An externally supplied gas may or may not be used.
- GMAW and FCAW processes are used in semiautomatic and automatic welding operations. In the semiautomatic mode, the operator controls the manipulation of the electrode. In the automatic mode, all weld variables are controlled by a machine or robot.
- Basic equipment requirements for GMAW and FCAW include a power source with constant voltage, DCRP capabilities, a wire feeder to supply the wire electrode, a gas supply (not needed for FCAW-S), a welding gun to direct the electrode, and a work clamp and weld cables to complete the electrical circuit.
- GMAW deposition modes are regulated by the amount of voltage and amperage (wire speed). The four modes are short-circuiting, globular, spray transfer, and pulsed spray transfer. Each has its own advantages and disadvantages.
- The FCAW process types are determined by the use of an external shielding gas supply. The two flux cored arc welding processes are self–shielded (FCAW–S) and gas-shielded (FCAW-G). Each process has advantages for certain applications.

Review Questions

Answer the following questions using the information provided in this chapter.

1. In the GMAW process, metallic parts are fused when they are melted by a(n) _____ between a continuously fed welding wire electrode and the workpiece.
2. The electrode used in the _____ process is a tubular flux-filled wire.
3. Shielding gas is used to prevent _____ of the molten metal during welding.
4. A constant _____ power source is used for GMAW and FCAW.
5. In a(n) _____ welding operation, the welder uses a handheld gun to deposit the weld metal.
6. List the four main deposition modes of the GMAW process.
7. The _____ deposition mode uses relatively low voltages and amperages.
8. In the _____ deposition mode, a drop of molten filler metal is melted off from the electrode wire at a controlled rate and time in the weld cycle.
9. In the self-shielded flux cored arc welding process, how is the molten weld metal shielded?
10. Two of the gases used for FCAW-G are _____ and _____.

Chapter
Welding Safety

Objectives

After studying this chapter, you will be able to:

❑ Identify the standard for safety in welding, cutting, and allied processes.

❑ Understand the requirements for personal protective equipment.

❑ Take necessary precautions when working with electrical current.

❑ Follow safety rules that apply to working with shielding gases.

❑ Maintain a safe welding environment.

Technical Terms

ANSI Z49.1
Dewar flasks
personal protective equipment (PPE)
phosgene gas
safety data sheets (SDS)

Welding Safety Standard

Welding can be performed safely with minimum risk if the welder uses common sense and follows safety rules. The American Welding Society (AWS) publishes *ANSI Z49.1 Safety in Welding, Cutting, and Allied Processes*. This national standard outlines welding safety practices for the protection of personnel and property in a detailed manner. ANSI Z49.1 is available free of charge from AWS. It is recommended that anyone involved in welding operations thoroughly understand the contents of this publication.

Establish good safety habits as you work with welding processes. Check equipment regularly and be sure that your environment is safe. Welders are responsible for protecting themselves from the conditions created during the welding process. This chapter covers three major safety areas—personal safety, welding process safety, and welding environment safety.

Personal Safety

Personal protective equipment (PPE) is necessary whenever an individual is in or around a welding environment. This equipment protects welders and others in the area from the sparks, heat, and the arc created during welding. PPE is also required for welders who are performing material preparation operations, such as cutting and grinding.

A good check of proper PPE is to start with the head and work down to the feet:

• **Head.** Welding cap, hard hat (if required), welding helmet, safety glasses or goggles, hearing protection, face shield, respiratory protective equipment (if required). See **Figure 2-1**.

- **Body.** Cotton, wool, or leather welding shirt, sleeves, or jacket; leather gloves; heavy cotton pants, such as jeans; leather apron.
- **Feet.** Leather boots with nonslip soles and steel toes (if required).

The welding helmet, cutting goggles, and face shield include a shaded lens. This lens is required to protect the welder's eyes from ultraviolet (UV) and infrared (IR) rays created during welding and cutting processes. As the shade number increases, the lens gets progressively darker. (See the Guide for Shade Numbers in the *Reference Section* of this book.) A clear plastic cover lens is installed over the shaded lens to protect that lens from scratches and breakage. Wear clear safety glasses or shields for eye protection when grinding, chipping, and power brushing. Eye protection should be worn at all times in a manufacturing or construction environment.

Hearing protection, such as earplugs or muffs, is recommended to protect against the noise levels often found in manufacturing operations. The hearing protection also keeps sparks and welding debris out of your ears.

All welding operations create fumes and smoke. Fume extraction systems and increased airflow should be used during welding and grinding operations. Fumes resulting from certain welding procedures may be highly toxic, and respiratory protective equipment must be worn. This equipment may also be required for welders working in confined spaces where oxygen may be reduced. The type of work being done determines the specific type of respiratory protection required.

Protective clothing should be made of a flame-resistant material, such as cotton, wool, or leather. There should be no rips, tears, or frayed material. This clothing is necessary to reduce the chance of cuts and burns. Welding gloves should fit comfortably and have the proper insulation for the desired weld process. For example, higher amperages are normally used for FCAW than for GMAW, thus requiring heavier gloves to resist the additional heat created. See **Figure 2-2**. It is important to remember that clothing designed to protect the welder can also cause the welder to

Goodheart-Willcox Publisher

Figure 2-1. PPE for the head includes a hard hat, eye protection, and hearing protection. This worker is also wearing a harness to protect against falls.

Goodheart-Willcox Publisher

Figure 2-2. This welder is wearing the proper welding PPE.

become overheated in warm environments. Precautions to avoid heat stress include drinking plenty of water, proper ventilation, regular breaks, and breathable clothing that wicks away moisture.

Welding Process Safety

The welding equipment used for GMAW/FCAW processes consists of the electrical power source and shielding gas cylinders. Both of these require proper care and handling for safe operation.

Electrical Current Safety

Primary voltage to the electrically powered welding machine is usually 115 V AC to 440 V DC, and in some cases may be higher. This voltage may cause extreme shock to the body and possibly death. For this reason, observe the following safety rules:

- Never install fuses of higher amperage than specified on the data label or in the operation manual.
- Install electrical components in compliance with all electrical codes, rules, and regulations. See **Figure 2-3**.

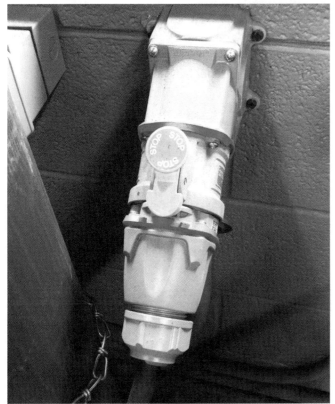

Goodheart-Willcox Publisher

Figure 2-3. This electrical outlet is properly installed for 440 V 3-phase power.

- Be certain that all electrical connections are tight.
- Never open a welding machine cabinet when the machine is operating. This should only be done by a competent electronic technician who is fully trained in machine installation and operation.
- Always lock primary voltage switches open and remove fuses when working on electrical components inside the welding machine.

Welding current supplied by constant current and constant potential power sources has an open circuit voltage range of 40 V to 80 V. At this low voltage, the possibility of lethal shock is small. However, this voltage still produces a shock that may cause health problems. To reduce the possibility of electric shock, do the following:

- Keep the welding power supply, power cable, work cable, and welding gun dry.
- Make sure the work clamp is securely attached to the power supply and workpiece.

Shielding Gas Safety

The gases used in GMAW/FCAW are produced and distributed to the user in either liquid or gaseous form. All storage vessels used for these gases are approved by the Department of Transportation (DOT), or previously by the Interstate Commerce Commission (ICC), and are so stamped on the vessel's cylinder walls. See **Figure 2-4**.

Some of the gases used in GMAW/FCAW are inert, odorless, and colorless. Therefore, special precautions must be taken when using them. None of the gases are toxic, but they can cause asphyxiation (suffocation)

Goodheart-Willcox Publisher

Figure 2-4. This cylinder has been made and tested to a DOT (Department of Transportation) specification. The name imprinted is the owner.

in a confined area without sufficient ventilation. An atmosphere that does not contain at least 18% oxygen can cause dizziness, unconsciousness, or even death.

Shielding gases cannot be detected by the human senses and can be inhaled like air. Never enter any tank, pit, or vessel where gases may be present until the area is purged (cleaned) with air and checked for oxygen content. High-pressure gas cylinders contain gases under very high pressure (approximately 2000 psi to 4000 psi) and must be handled with extreme care.

Observe the following rules:
- Store all cylinders in a vertical position.
- Secure all cylinders with safety chains, cables, or straps. See **Figure 2-5**.
- Know the contents before use. **Figure 2-6** illustrates a label indicating the type of gas within the cylinder and safety precautions for shipping and using the gas cylinder.
- Keep the protective cap in place until the cylinder is ready to use.

- Do not move a cylinder without the protective cap in place. Always use a cylinder cart, with the safety chains installed, to move a cylinder.
- Check the outlet threads and clean the valve opening by cracking (slightly opening and closing) the cylinder valves before attaching the regulator. See **Figure 2-7**.
- Use the proper equipment for the type of gas used.
- Attach the regulator and securely mount the flowmeter in the vertical position, **Figure 2-8**.

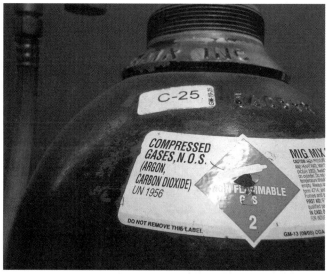

Figure 2-6. The cylinder label indicates the type or mixture of gas.

Figure 2-7. Cracking open the cylinder before attaching the regulator/flowmeter blows out any dirt in the valve opening.

Figure 2-5. Secure all cylinders with chains when moving or storing them.

Figure 2-8. This regulator/flowmeter has been properly attached in a vertical position.

- Before opening a cylinder valve, ensure the regulator adjusting screw is fully backed out and free of tension.
- Stand aside and slowly open the cylinder valve. Never stand in front of the gauges when opening a cylinder valve.
- When cylinders are not in use, the cylinder valve should be closed and the regulator adjusting screw should be fully loosened.
- When a cylinder is empty, close the valve, replace the safety cap, and mark the letters *MT* with soapstone on the upper part of the cylinder. See **Figure 2-9**.

Figure 2-9. An empty cylinder marked with *MT*.

- Never tamper with a leaky valve—return the cylinder to the supplier for a replacement.

Liquefied gas cylinders, commonly called *Dewar flasks*, are basically vacuum bottles. See **Figure 2-10**. The gas has been reduced to a liquid at the supplier's plant for ease in handling and storage. The liquid is converted to a gas for welding by heat exchangers within the cylinder or as a part of the gas delivery system in the welding facility. The following safety rules apply to Dewar flask systems:

- Always keep cylinders in the vertical position.
- Move cylinders on cylinder carts. These cylinders are extremely heavy and difficult to handle.
- Always use the proper equipment when installing or connecting cylinders.
- Do not interchange equipment components.
- Liquid gases are extremely cold and cause severe frostbite if they come in contact with the eyes or skin. Always wear gloves and safety glasses. Do not touch any frosted surface with bare hands.

Figure 2-10. Liquefied gas cylinders are commonly called *Dewar flasks*.

Welding Environment Safety

It is the responsibility of every welder to be aware of the environmental hazards encountered on the factory floor or around a construction project. Pay close attention to the following rules to help maintain a safe welding environment:

- Keep the welding area clean.
- Keep combustibles out of the welding area.
- Maintain good ventilation in the weld area.
- Repair or replace worn or frayed ground or power cables.
- Make sure the part to be welded is securely grounded.
- Make sure welding helmets have no light leaks.
- Use a proper shade number lens to protect the eyes from arc radiation. Shade darkness is determined by the welding amperage being used.
- Wear tinted safety glasses when others are tack welding or welding near you.
- Use safety screens or shields to protect your work area.
- When welding on cadmium-coated steels, copper, or beryllium copper, use special ventilation to remove fumes and vapors from the work area. See **Figure 2-11**. Portable smoke

Goodheart-Willcox Publisher

Figure 2-11. This welder is using a fume exhaust system to remove toxic smoke and fumes.

removal systems are also available for this purpose.

- Do not weld near trichloroethylene vapor degreasers. The arc changes the vapor to poisonous *phosgene gas*, which is odorless and colorless. A sweet taste in your mouth indicates that this gas is being formed.
- Fires may be started by welders in a number of ways, such as igniting combustible materials, misuse of fuel gases, electrical short circuits, and improper storage of hot material. Know where fire extinguishers are located and their proper use and limitations. See **Figure 2-12**. Having another person act as a fire watch is highly recommended. If a fire becomes out of control, leave the area and contact the proper authorities.
- Never weld on a container that has previously held a fuel until you are sure that it has been purged with an inert gas and tested for fume content.
- Never enter a vessel or confined space that has been purged with an inert gas until the space is checked with an oxygen analyzer to determine that sufficient oxygen is present.
- Never use oxygen in place of compressed air. Oxygen supports combustion and makes a fire burn violently.
- Always wear safety glasses or safety shields when using power brushes. This equipment is dangerous and can eject broken pieces of wire.
- Pay close attention to the clamping operation when working with mechanical, hydraulic, or air clamps on tools, jigs, and fixtures. Serious injury can result if parts of the body are caught in the clamp.
- Be aware of moving equipment, such as forklifts and vehicles.
- Be aware around robotic equipment and follow safe space requirements.
- Always be aware of what is happening in your surrounding area.
- Inform others before you begin any potentially dangerous activity.

Safety Data Sheets and Labeling of Chemicals

OSHA's Hazard Communication Standard (HCS) requires employers to provide employees with information about the health and physical hazards of substances in the workplace. Employers must ensure

Fires	Type	Use		Operation
Class A Fires Ordinary Combustibles (Materials such as wood, paper, textiles.) *Requires...* *cooling-quenching.* Old New	**Soda-acid** Bicarbonate of soda solution and sulfuric acid	Okay for use on A		Direct stream at base of flame.
		Not for use on B C D		
Class B Fires Flammable Liquids (Liquids such as grease, gasoline, oils, and paints.) *Requires...blanketing or* *smothering.* Old New B	**Pressurized Water** Water under pressure	Okay for use on A		Direct stream at base of flame.
		Not for use on B C D		
	Carbon Dioxide (CO$_2$) Carbon dioxide (CO$_2$) gas under pressure	Okay for use on B C		Direct discharge as close to fire as possible, first at edge of flames and gradually forward and upward.
		Not for use on A D		
Class C Fires Electrical Equipment (Motors, switches, etc.) *Requires...* *a nonconducting agent.* Old New C	**Foam** Solution of aluminum sulfate and bicarbonate of soda	Okay for use on A B		Direct stream into the burning material or liquid. Allow foam to fall lightly on fire.
		Not for use on C D		
	Dry Chemical	Multi-purpose type	Ordinary BC type	Direct stream at base of flames. Use rapid left- to-right motion toward flames.
Class D Fires Combustible Metals (Flammable metals such as magnesium and lithium.) *Requires...blanketing or* *smothering.* Old New		Okay for A B C	Okay for B C	
		Not okay for D	Not okay for A D	
	Dry Chemical Granular type material	Okay for use on D		Smother flames by scooping granular material from bucket onto burning metal.
		Not for use on A B C		

Goodheart-Willcox Publisher

Figure 2-12. Fire extinguisher classification chart.

that all containers are properly labeled, employees are provided access to safety data sheets, and a training program is conducted for all employees who may potentially be exposed to the chemicals. Welders work with chemicals such as degreasers and solvents.

Safety Data Sheets (SDS) detail the properties and hazards of chemical products. These sheets include precautions such as PPE to be worn and the handling, storage, and disposal of substances. Safety data sheets describe the product and include information about its hazardous components, flammability, reactivity, and many other important characteristics pertaining to the safe use of the product.

Chemical manufacturers use labels that include a signal word, pictogram, and hazard statement. See **Figure 2-13**. Two signal words are used—*"Danger"* and *"Warning."* *Danger* is used for the more severe hazards, and *Warning* is used for the less severe hazards. Only one signal word is used on a label no matter how many hazards a chemical has. If the *Danger* signal word applies to one hazard and *Warning* applies to another, only *Danger* will appear on the label.

Hazard Communication Standard Pictograms and Hazards

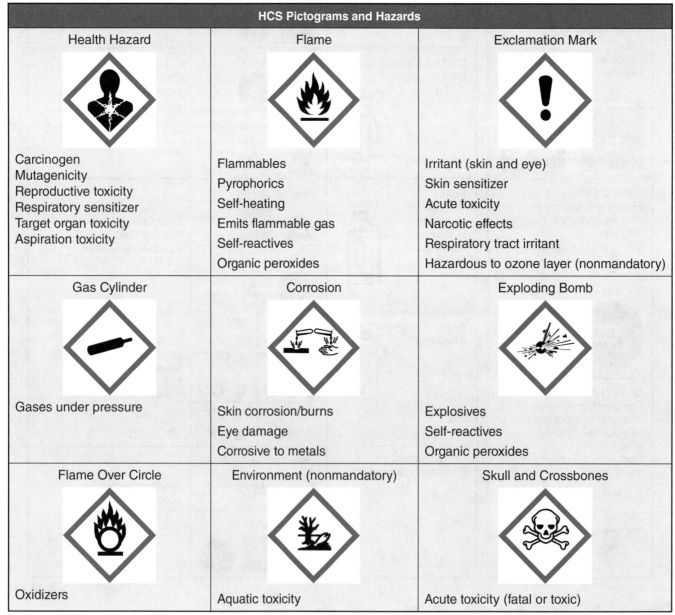

US Department of Labor

Figure 2-13. HCS pictograms and hazards.

Summary

- The standard for safety in welding, cutting, and allied processes is ANSI Z49.1. This document is available free of charge from the American Welding Society.
- Welding safety includes personal safety, welding process safety, and welding environment safety.
- Personal protective equipment (PPE) is essential to protect those in or around a welding environment from the sparks, heat, and arc created during welding.
- The primary voltage to a welding machine can cause extreme shock to the body and possibly death. Working safely with this equipment requires strict observance of electrical safety rules.
- Inert, odorless, and colorless shielding gases can cause suffocation in a confined area without sufficient ventilation.
- High-pressure gas cylinders contain gases under very high pressure and must be properly handled and stored for safety.
- The many environmental hazards associated with welding operations need to be recognized and the appropriate precautions taken. Awareness of your surroundings and respect for dangerous conditions are required at all times.

Review Questions

Answer the following questions using the information provided in this chapter.

1. The standard for safety in welding, cutting, and allied processes is _____.
2. What are the three main safety areas described in this chapter?
3. What type of footwear should be worn during welding?
4. List three flame-resistant materials used for protective clothing.
5. The voltage supplied to an electrically powered welding machine is usually _____ V AC to _____ V AC.
6. *True or False?* Shielding gases can be detected by their odor.
7. *True or False?* All cylinders should be stored in a vertical position.
8. When a cylinder is empty, you should close the valve, replace the safety cap, and mark the letters _____ with chalk on the upper part of the cylinder.
9. Liquefied gas cylinders are commonly called _____ and are basically vacuum bottles.
10. List three ways a fire might be started in a welding area.
11. To protect welders from fumes and smoke, _____ systems and increased airflow should be used during welding and grinding operations.
12. Wearing the proper shade number welding or cutting lens protects the eyes from what two types of rays?
13. Liquid gases cause severe _____ if they come in contact with the skin.
14. Welders should not weld on a container that has previously held a fuel or chemical until they are sure it has been purged with a(n) _____ gas and is free of fumes.
15. *True or False?* Oxygen can be used in place of compressed air.

Personal protective equipment is required to protect welders from the sparks, heat, and arc created by welding.

Chapter 3

Equipment Setup and Controls

Learning Objectives

After studying this chapter, you will be able to:
- ❏ Identify power source controls and their functions.
- ❏ Describe proper power source installation and maintenance.
- ❏ Distinguish three types of electrode feeders.
- ❏ Identify feeder controls and their functions.
- ❏ Explain the functions of cables and guns.

Technical Terms

arc voltage
contact tips
duty cycle
inductance
nozzles
open circuit voltage
push-pull type feeder
push-type feeder
spool gun

Introduction to Equipment Setup

The basic power sources and equipment for the GMAW and FCAW processes are essentially the same. Both processes use a constant voltage power source. This chapter describes the equipment components, along with the functions and adjustments required for efficient operation.

Several variables must be considered before adjusting welding equipment, including the process, base material, electrode, shielding gas, and desired weld. These variables are discussed separately in future chapters, yet it is important to note that a change in any one variable requires adjustments to the others and to the equipment setup. Correct setup of equipment for GMAW and FCAW processes is essential. Proper setup guarantees intended performance standards and ensures that welding can be done safely without equipment malfunctions.

Power Sources and Specifications

A power source produces current at a low voltage for melting the electrode. The equipment controls the welding operation in several areas:
- Input voltage (primary voltage).
- Open circuit voltage.
- Output ratings and performance.
- Duty cycle.

The National Electrical Manufacturers Association (NEMA) has established specification *EW-1 Electric Arc Welding Power Sources* for control of these areas. Each type of power source is designed for a specific purpose, with limitations established to ensure proper operation. Specifications for machines using utility power fall into the following categories:

- **Primary power type, voltages, and cycles.** Includes alternating current, single-, or three-phase power at 110 V, 208 V, 230 V, and 460 V and 60 hertz (Hz) cycles.
- **Primary power fusing.** Fuse sizes are specified for individual machines. Limits must not be exceeded.
- **Rated welding amperes.** Current amounts specified by the manufacturer should not be exceeded since the cooling system is not capable of carrying away excess heat.
- **Duty cycle.** All welding power sources, as well as cables, guns, and work clamps, are designed to operate for a specific time period at a specific load. *Duty cycle* is the percentage of time in a 10-minute period that a machine can run at its rated output without overheating. Duty cycle depends on the following:
  - Size of internal wiring.
  - Type of internal components.
  - Insulation of internal components.
  - Amount of cooling required.

The NEMA specification establishes that every 10% of duty cycle represents one minute of operation in a 10-minute period. A 150 ampere (A) machine, for example, would be rated a 30% duty cycle if it can operate continuously for three minutes at 150 A without overheating. Before welding could be resumed, the machine then must be idle seven minutes to allow internal components to cool.

Duty cycles range from 20% to 100%. Electric arc welding machines are classified as follows:

- Class I—60%, 80%, or 100% duty cycle; industrial use machines.
- Class II—30%, 40%, or 50% duty cycle; light industrial use machines.
- Class III—20% duty cycle; home and hobby use machines.

Welding equipment not made to NEMA specifications has duty cycles specified by the manufacturers. Never exceed an equipment manufacturer's duty cycle requirements. See **Figure 3-1**.

Duty Cycle	Number of Minutes Machine Can Be Operated at Rated Load in a 10-Minute Period
100%	Full Time
60%	6
50%	5
40%	4
30%	3
20%	2

Goodheart-Willcox Publisher

Figure 3-1. Power source duty cycles limit the number of minutes that a unit can operate at the rated load without overheating.

Constant Voltage DC Power Source Controls

The type and number of controls vary with the power source. They range from a tap connection and a simple rheostat to numerous controls necessary for high-quality welds. The controls found on a typical power source are discussed in the following section.

Open Circuit Voltage (OCV) Control

Open circuit voltage is the voltage output of the power source before an arc is established. The output values are established by the power source manufacturer. The range of voltages is adjusted with taps, levers, or switches. The basic response output of a power source can be computed on the machine volt-ampere curve. A constant voltage power source does not have true constant voltage output. It has a slight downward or negative slope due to internal electrical resistance in the welding circuits, causing a minor droop in the output volt-ampere characteristics.

When setting up a machine with a voltmeter on the panel, follow these steps to establish the proper OCV range for the arc voltage to be used:

1. Turn the power source on.
2. Release the wire feeder idler roller pressure to prevent feeding.
3. Place the voltage range switch in the desired location. Check the manual for OCV and the arc voltage range to select the range of values.
4. Hold the gun away from the work clamp or workpiece and energize the contactor switch on the gun. When you depress the switch, the contactor allows electrical current to flow to the electrode tip and the welding electrode. An arc will form if the electrode contacts the work.

5. Observe the voltmeter and adjust the fine-tuning voltage control to the desired OCV. The arc voltage will be 2–3 volts lower for each 100 amperes.

6. Release the trigger or switch and reset the tension on the idler roller.

The voltage output of the power source when an arc is present is called the *arc voltage*. Adjust the fine-tuning control (the same control used to set the OCV) to the desired voltage. This control can be adjusted during welding.

Slope refers to the slant of the volt-ampere curve and the operating characteristics of the power source under load. In many machines, the slant of the volt-ampere curve is automatically set as the OCV is changed. In other machines, the curve can be adjusted for various welding conditions. FCAW operates best with the power source set in the flat mode.

Inductance Control

Inductance is the separation of the molten drops of metal from the electrode controlled by squeezing forces exerted on the electrode due to the current flowing through it. See **Figure 3-2**.

Some power sources have an inductance control on the main panel. However, this control has no effect

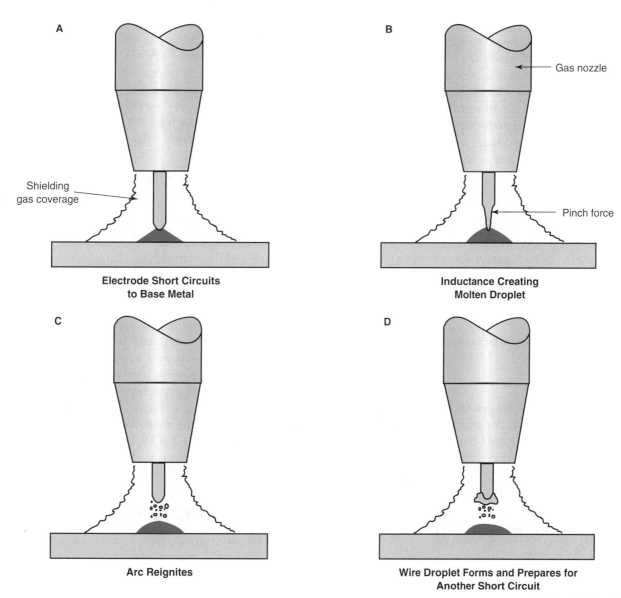

Goodheart-Willcox Publisher

Figure 3-2. Short circuiting transfer sequence. A—The electrode wire short circuits to the base metal and creates a weld pool. B—The pinch effect squeezes the molten droplet off and into the weld pool. C—The welding arc reignites. D—The electrode nears another short circuit and the process begins again.

on the way the electrode wire melts in the FCAW operation. Inductance control is primarily used for GMAW with short circuiting transfer.

By adjusting the level of inductance, the machine controls the rate of current rise when the electrode contacts the workpiece and the short circuit starts. If the current flows rapidly through the electrode, the drop of metal is squeezed off quickly and causes spatter. If inductance is added to the circuit and the current is not applied as rapidly, the number of short circuits per second decreases and the "arc on" time increases. This results in a more fluid weld pool, a smoother weld crown, and less spatter.

In GMAW spray transfer mode, some inductance is beneficial in starting the arc process. This limits explosive starts by slowing down the rate of current rise at the start of the cycle.

As a general rule, both the amount of short-circuit current and the ideal amount of inductance increase as the electrode diameter increases. The type of electrode to be used also affects the inductance setting. A harder material, such as stainless steel, requires a higher inductance to increase the fluidity of the weld pool. The finished weld bead will appear flatter and smoother with an increase in inductance. A softer material, such as aluminum, welds better with a lower inductance. The lower inductance reduces some of the fluidity in the weld pool. The following is a useful guide for setting up the machine for welding in the GMAW short circuiting mode:

Maximum inductance produces these results:
- More penetration.
- More fluid weld pool.
- Flatter weld.
- Smoother bead.
  Minimum inductance produces these results:
- More convex bead.
- Increased spatter.
- Colder arc.

The function of the power source is to maintain a preset voltage condition. The feeder produces a continuous electrode feed and, when required, the machine provides more or less welding current to maintain the preset arc length. Thus, the machine is called a *constant voltage* or *constant potential* power source and produces only direct current.

Preprogrammed Welding Machine Controls

When using a power source that has built-in programs for various types and thicknesses of materials, select the program desired and insert the required procedure data. This automatically sets the correct machine values for the operation and monitors the machine output during the welding operation. The face of the machine has push buttons or dials for program selection and modification of each value desired. See **Figure 3-3**. Each readout of the program is displayed on the digital meter. The factory program can also be modified by manual override, if desired.

Each machine also has areas where programs can be installed and modified by the operator for individual welding procedures. Machines of this type are usually factory programmed for the welding of steel, stainless steel, and silicon bronze or aluminum.

Pulsers and Auxiliary Pulsers

Some machines have a pulsing system in the main power source. If the power source does not contain a pulse setting, an auxiliary machine can be added for pulsing the welding operation. Units are attached to the system and can be manually set for the desired pulse on and off times. See **Figure 3-4**. Always refer to the operation manual for a list of setup options for this process mode and applications.

Constant Current Power Source Controls

A constant current (CC) power source can be used for FCAW with a special voltage sensing wire feeder system. A constant current machine is designed to have the primary adjustment for the amperage. The voltage is determined by the arc length. This power source is used for the shielded metal arc welding (SMAW) and gas tungsten arc welding (GTAW) processes.

A CC machine requires an OCV slope set as flat as possible. The only way to control slope is to place the amperage range switch (if available) in the highest position. The main amperage rheostat is set for the amount of welding current desired to burn off the electrode fed into the molten pool. When a CC power source is used for GMAW, the voltage sensing wire feeder is also required (see the *Wire Feeders* section).

Power Source Installation

A power source should be installed in an area free of dust, dirt, fumes, and moisture. Machine heat must be able to escape. Both dirt and improper cooling can cause a power source to overheat and the internal components to be ruined. Electronic components that absorb moisture can fail. Make sure objects do not

Wirefeed speed or amperage display

Wirefeed adjustment knob

Main display knob adjusts program information

2 stop/4 stop button for trigger operation

On/off switch

Voltage or trim display

Voltage or trim adjustment knob

Main display shows program information

Options button

Procedure memory buttons

Goodheart-Willcox Publisher

Figure 3-3. This power source has preinstalled programs that can be modified as needed and saved for future use. All readouts are digital.

Used with permission of The Lincoln Electric Company, Cleveland, Ohio, U.S.A.

Figure 3-4. This auxiliary unit attached to the power source provides pulsing and additional advanced technology process capabilities.

block the free flow of air into or out of the machine. Most machine manufacturers specify the required space for air circulation around a power source.

Power sources use various input voltages and 60 Hz cycles. Machines using other than 60 Hz are specially made. The required fuse size for the incoming power is shown on the data label, **Figure 3-5**. The fuse

THE LINCOLN ELECTRIC CO. CLEVELAND, OHIO U.S.A.		POWER WAVE™ C300		
CODE - SERIAL NO. N° DE CODE - N° DE SERIE NO. CODIGO - NO. DE SERIE	11479	U1090904956		
(1) 3~ 🔲⬇️⬜⬛ 〰️		IEC 60974-1 GB15579.1-2004		

		⎓	X	40%	100%
⚡	55A/22.2V TO 280A/31.2V	U₀ = 80V	I₂	280 A	225 A
			U₂	31.2V	29 V
⚡	5A/10.2V TO 300A/22V	U₀ = 80V	I₂	300 A	250 A
			U₂	22 V	20 V
⚡ ⚡	40A/16V TO 300A/29V	U₀ = 80V	I₂	300 A	250 A
			U₂	29 V	26.5 V

	U₁		I₁ MAX		I₁ EFF	
		1	3	1	3
🔌	1~ OR 3~ 50/60 HZ	208V	53A	30A	41A	23A
		220/230V	48A	28A	37A	21A
		380/400/415V	29A	16A	22A	12A
		460V	25A	14A	19A	11A
		575V	20A	11A	16A	9A

IP 23	INS.CL.155(F)	ⓒ		Ⓢ	
					S27260 VM

Goodheart-Willcox Publisher

Figure 3-5. The data label found on all power sources indicates the required fuse size for input power, along with duty cycle ratings.

panels should always be close to the machine so the main power can be disconnected in an emergency. Machines that do not match the utility power voltage can be used with a step-up or step-down transformer.

Power Source Maintenance

Insulated transformers, solid-state components, and sound designs have extended the life of modern welding power sources. Given reasonable care and routine maintenance, a welding power source will operate satisfactorily for many hours before repairs are needed. When performing maintenance or repair of a power source, follow the manufacturer's instructions and these guidelines:

- Turn off the machine and disconnect the circuit at the fuse box before beginning maintenance.
- Clean or blow out the unit periodically. Use only dry, filtered, compressed air, nitrogen gas, or an electrical (nonconducting) cleaner. Always wear a face shield when using compressed air or gas.
- Check all terminals for loose connections.
- Lubricate fan and motor bearing, if required.
- Check mechanical arms and switches for freedom of movement. Lightly grease mechanical connections.
- Clean terminal blocks and tap connectors with a wire brush or abrasive pad. If corroded, they will restrict the flow of electrical current. Make sure terminal blocks and tap connectors are tight.
- Check motor generator brushes and replace them when they are worn beyond the manufacturer's tolerances. Worn brushes wear armatures, which must then be machined for proper operation.
- Lubricate, inspect, and adjust portable gasoline and diesel power sources. Improper maintenance reduces their capacity and operation.
- Use only authorized replacement parts.

Wire Feeders

Most feeders use 110 V AC power, which is provided to the machine by a connection in the power source. If this connection is used, the wire feeder is turned on when the power source is operating and turned off when the power source is not operating. Wire feeder systems may be included as part of the power source or as a separate unit connected by control cables. It is important to note the direction the wire feeds from the spool. Depending on the type of feeder, the wire may spool from either the top or the bottom of the roll and

should be installed accordingly. Wire feeders that are installed on push-pull systems or in handheld spool guns use 24 V DC motors for precision drive of the wire and for the safety of the welder.

A portable semiautomatic unit is used for welding at a distance from the basic power source and can be used in either a CC or CV mode. In the CV mode, the unit is set up as a standard feeder. In the CC mode, the unit uses a voltage-sensing motor to drive the electrode at the proper speed. Before welding begins, the speed and amperage relationship must be established from a setup chart. The correct polarity is established on the feeder, and a voltage-sensing clip is installed on the workpiece.

The voltage adjustment will not operate if the unit has a voltage-adjustment dial on the face of the feeder and the power source is CC. The arc voltage must be set using a setup chart. Unless a special circuit has been installed on the drive unit, the electrode will be electrically "hot" (charged) when connected to an operating power source.

Wire Feeding Systems

The basic types of wire feeding systems used for GMAW/FCAW are *push-type feeders*, *push-pull type feeders*, and *spool guns*:

- **Push type.** One or two sets of feed drive rollers push the wire from the spool to the welding gun. See **Figure 3-6**.
- **Push–pull type.** One set of feed drive rollers pushes the wire from the wire feeder to another set of drive rollers mounted in the welding gun. This system is used for welding with soft or small-diameter wires, such as aluminum, since these wires may buckle in the cable liner if pushed long distances. See **Figure 3-7**.
- **Spool gun type.** This unit has a spool of wire located on the welding gun. Since the size of the wire supply spool is small, the amount of welding is limited. See **Figure 3-8**.

Wire Feeder Controls

Controls vary on different wire feeders and wire feeders are selected based on the intended use and amount of wire feed desired. See **Figure 3-9**. The major controls include the power switch (off/on control), the wire feed speed (potentiometer) control, and the spool brake tension (stops wire spool at end of welding) control.

Auxiliary wire feeder controls may include the following:

- **Mode switch.** Used to select various modes, such as spot, intermittent, and seam welding.

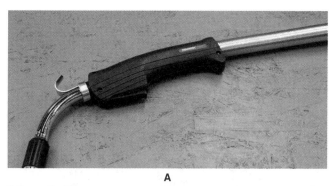

A

Driven roll with
gear behind it

Driving roll with
gear behind it

B

Goodheart-Willcox Publisher

Figure 3-6. Push type wire feeding system. A—This gun is for use with a push-type wire feeder. B—A set of wire feeder drive rollers used for both types of wire feeder systems.

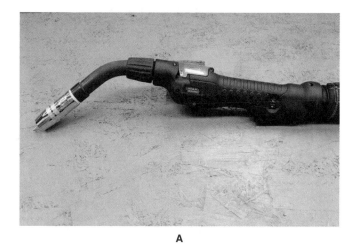

A

Figure 3-7. Push-pull wire feeding system. A—This gun has a pull-type wire feeder built in. Because it is used in conjunction with a push-type feeder, it is called a push-pull gun. B—A push-pull wire feeder system includes a separate set of drive rolls in the gun handle.

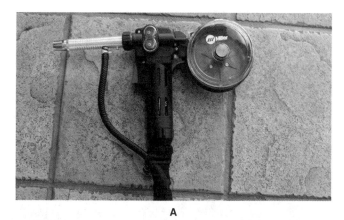

A

B

Goodheart-Willcox Publisher

Figure 3-8. Spool gun. A—A spool gun feeder system uses small diameter rolls of electrode wire that limit welding capabilities. B—A standard two-pound roll of electrode wire for use in a spool gun feeder system.

B

Goodheart-Willcox Publisher

Goodheart-Willcox Publisher

Figure 3-9. The number and types of controls included in wire feeders vary.

Brake adjuster

Goodheart-Willcox Publisher

Figure 3-10. The brake adjuster is a tension device located in the hub of the wire spool.

- **Trigger two-stage lock control.** Allows the welder to weld without the need to depress the operating switch during the entire operation.
- **Burnback control.** Prevents the wire from sticking in the weld pool. It sets the time that the arc power is on after the stop switch is released to burn back the wire from the molten weld pool. This control must be used for spot welding.
- **Spot-weld time control.** Sets the length of time for the spot weld operation.
- **Stitch weld timer.** Sets the time needed to make a long seam or a specific weld length.
- **Pre-purge timer control.** Establishes the time that the gas flows before welding will start.
- **Purge control.** Allows the welder to open the gas solenoid. The amount of gas flow can be set on the flowmeter, or the welder can purge the gas lines prior to welding.
- **Jog control.** Moves the wire through the wire feeder to the contact tip. This control is used when a new wire is fed to the gun. When determining wire speed, the welder may also use the jog control. (Jog speed is the same speed as the set wire speed.)
- **Wire reel brake adjustment.** A tension device in the hub of the wire spool that is adjusted so the spool stops rotating at the same time the wire stops feeding at the end of the welding operation. If the wire is allowed to overspool and touch the feeder housing, a short circuit can occur if the electrode circuit is energized or the wire may birdnest (tangle). When changing reels of wire, always replace the reel locking pin and check the brake adjustment. See **Figure 3-10.**

Wire Straightener

Wire straighteners are placed in the supply system to reduce the amount of wire cast (curvature) before the wire is drawn into the drive roller system. Excessive cast is often found on hard wires, causing heavy wear of the contact tips during welding. This can result in arc outages (loss of arc). The amount of cast is increased or decreased with an adjusting wheel on the straightener.

Wire Feeder Drive Roll System

The system for driving the electrode wire from the spool to the gun is either a two-roll or a four-roll system. See **Figure 3-11** and **Figure 3-12**. Several

Goodheart-Willcox Publisher

Figure 3-11. A two-roll drive system is standard on most wire feeders.

Figure 3-12. A four-roll drive system is ideal for large-diameter electrode wire and hard wires like stainless steel.

different types of drive rollers are shown in **Figure 3-13**. Some drive rollers are grooved for driving specific sizes and types of wire. Drive rolls may be knurled, grooved, or cogged in *V* or *U* shapes. The *V* grooved roll is used for hard wire GMAW and hard-casing-type (Innershield®) electrodes for FCAW-G. The *U* grooved roll is used for soft wires such as aluminum GMAW and soft-casing-type (Outershield®) electrodes for FCAW-S. Serrated drive rolls are not used in the FCAW process. If the groove does not match the wire type and size, the wire will seize in the groove and not feed properly.

The idler roller part of the system may be smooth or have grooves for the specific wire type and

diameter. The idler roller must be adjusted to feed the wire into the roller grooves without slippage or flattening the wire.

Inlet guides are used to feed the wire into the drive rollers. Outlet guides are used to feed the wire from the drive rollers into the gun cable or cable adapter. Guides are made of a variety of materials. Hard wires (such as steel, stainless steel, and Inconel) wear the guides quite rapidly and require frequent replacement. Guides are adjusted as close to the rollers as possible to prevent the wire from birdnesting or balling up in the area between the rollers and guides. See **Figure 3-14**.

Wire Wipers

This accessory is attached to the inlet side of the drive roll system, as shown in **Figure 3-15**. A wire wiper usually consists of a piece of felt material that is closed where the wire passes through it. A small amount of commercial fluid may be placed on the felt to lubricate the wire as it passes through. This lubrication extends the life of the guides and the cable liners in the gun cable. Use the wiper only when welding on the specific materials defined by the manufacturer of the fluid. Do not overlubricate.

Wire Feeder Maintenance

Wire feeders require very little maintenance to keep them operating properly. As with any electrical device, they must be kept dry to protect the electrical components within the feeder. Only a skilled electrician should check electrical malfunctions. The printed circuit boards (if the feeder is so equipped) should be

U-grooved

V-grooved

Knurled

Figure 3-13. Different types of wire feeder drive rollers.

Drive roll
tension adjustment

Drive
rolls

Wire guide
insert

Figure 3-14. Wire guides eliminate space in which the electrode wire can kink, ball up, or tangle (birdnest). This machine has a single guide assembly with integral inlet and outlet guides.

Figure 3-15. Wire wipers are used to clean the electrode before it feeds through the rollers and gun liner.

replaced only with factory-authorized parts obtained from the welding supplier.

Problems may occur in the drive roll system and the guides to and from the rollers. Worn equipment causes many problems in the system. Inspect the guides and rollers often and replace them when they are worn. Feeding problems are usually due to a problem in the wire feed mechanism. To determine if the feeder is at fault, disconnect the gun cable and run the wire through the feeder only. The wire should run through smoothly. If the feeder is operating satisfactorily, connect the gun and cable to the feeder. Operate the feeder with the gun and cable attached. If the fault remains, the problem is in the gun and cable assembly. Be sure you know where a problem is before you try to repair it.

Gun Cables

Depending on the features of the wire feeder and gun, the gun cable provides a number of passageways between the feeder and the gun. Different passageways can carry electrical current, shielding gas, the welding electrode, cooling water, and vacuum (for smoke removal). With all of these operations to perform, cables must be cared for properly. Keep gun cables as straight as possible. Protect them from falling objects and wheels running over them. Cables are available in various lengths. Some cables can be connected together to make longer assemblies.

Adapters are available for attaching different guns and feeders. Gun manufacturers make adapters for almost every type of feeder.

A liner is installed in a cable to protect it from wear. Specific liners are made for use with certain sizes of electrodes. Some wires require a special lining material, such as Teflon® or nylon. The cable liner is always installed into the gun cable from the feeder end and secured at the welding tip end, **Figure 3-16**. Always follow the manufacturer's installation instructions. Liners are not interchangeable. When changing spools and types of wire, always blow out the liner with clean, dry, compressed air to remove any residue from the previous wire.

Welding Guns

GMAW/FCAW welding guns are designed to allow the welder to control the delivery of the electrical current, the shielding gas, and the energized electrode to the weld area. Welding guns use three basic electrode-feeding systems. The push type uses the drive rollers in the wire feeder to push the wire through the gun to the contact tip. The push-pull system uses the drive rollers in the wire feeder to push the wire in conjunction with a set of drive rolls in the gun handle that pull the wire. Spool guns are designed for welding aluminum or other soft alloy wires. They contain a low voltage motor and drive rollers for moving the electrode from the spool to the contact tip. A rheostat and a trigger are used to start and end the sequence and control electrode speed. The drive and idler rollers are of various designs. For this reason, when changing types of wire, diameters, or replacing parts, the rollers must be of the same

model and manufacturer for the system to operate as designed.

All types of guns are rated for a specific duty cycle and may be cooled through various means. Gun cooling systems include air-cooled, gas-cooled, and water-cooled:

- **Air-cooled guns.** These guns are used only with a self-shielded FCAW electrode. The guns have a small nozzle on the end that does not need to be removed for welding. See **Figure 3-17**.
- **Gas-cooled guns.** These guns are used with the smaller diameter GMAW electrodes and gas-shielded FCAW electrodes. A gas-cooled gun has an adapter on the end for the gas nozzle. The nozzles are available in different sizes and are usually copper or chrome-plated. See **Figure 3-18**.
- **Water-cooled guns.** These guns are used with a large diameter electrode and have a high duty cycle at a high amperage rating. A water-cooled gun may have a water cooler mounted in a closed system. Most modern systems use a specialized coolant to reduce mineral contamination to the system.

The handle of the gun holds all of the electrical components and the trigger. The electrical components should fit together firmly within the handle and gun assembly for proper operation of the gun. See **Figure 3-19** and **Figure 3-20**. The gun is used to direct

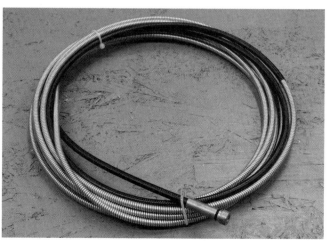

Goodheart-Willcox Publisher

Figure 3-16. A cable liner is installed into the gun cable from the feeder end and secured at the welding tip end.

Used with permission of The Lincoln Electric Company, Cleveland, Ohio, U.S.A.

Figure 3-17. Air-cooled guns, used only with a self-shielded FCAW electrode, have a small nozzle on the end that does not need to be removed for welding.

Used with permission of The Lincoln Electric Company, Cleveland, Ohio, U.S.A.

Figure 3-18. Gas-cooled guns, used with the smaller diameter GMAW electrodes and gas-shielded FCAW electrodes, have adapters on the end for the gas nozzle.

Goodheart-Willcox Publisher

Figure 3-19. One type of gun uses a screw-on insulator, gas diffuser, and contact tip with a slip-on gas nozzle.

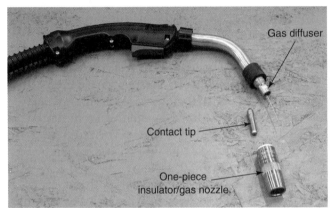

Gas diffuser

Contact tip

One-piece insulator/gas nozzle

Goodheart-Willcox Publisher

Figure 3-20. One type of gun uses a one-piece insulator/gas nozzle with a slip-in contact tip. The end of the cable liner is shown extending from the gas diffuser.

the electrode wire, transfer electrical current to the contact tip and the electrode, and evenly diffuse the shielding gas over the weld zone. When the trigger is held down, the gas solenoid is opened, and power is sent to the drive roll assembly and contact tip. Some manufacturers have combined some of these parts into a single unit to simplify the replacement process. Insulators that separate the welding current from the gas nozzle should be replaced if they are cracked or burned.

Contact Tips

Contact tips conduct electrical current from the power source to the consumable welding wire. Contact tips are designed for operation with a certain type and diameter of filler material and a specific transfer mode. Each transfer mode defines the relationship of the contact tip end with the end of the gas nozzle. As a general rule, short contact tips are used with spray transfer, longer contact tips are used with the short circuiting transfer, and heavy-duty contact tips are used for FCAW. See **Figure 3-21.**

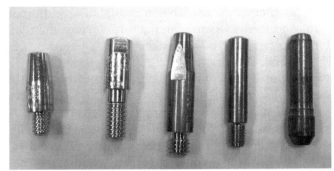

Goodheart-Willcox Publisher

Figure 3-21. Contact tips are determined by size of electrode wire, intended use and brand, style, and type of gun.

When the exit hole size becomes enlarged to the point where electrical contact with the wire cannot be maintained, the tip must be replaced. Continued use of an oversize hole in the contact tip causes arc outages, incorrect heating of the wire, incomplete fusion, and poor weld quality. Always use the manufacturer's recommended contact tip for the job involved and inspect it regularly to make sure the tip is seated firmly into the adapter seat. Since the tips are expendable, it is suggested that several tips for each application be kept in stock.

Gas Nozzles

Gas *nozzles* are tubes that attach to the end of the welding gun and direct the shielding gas flow over the weld area. Gas nozzles are made by each manufacturer to fit an individual gun, and they are not generally interchangeable. See **Figure 3-22.** However, manufacturers generally use the same system for dimensioning the exit size of the nozzle end. The exit diameter is usually dimensioned in sixteenths of an inch. In some cases where high currents are carried, the gas nozzle gets extremely hot, causing the copper to peel or flake. To correct this condition, some nozzles

Goodheart-Willcox Publisher

Figure 3-22. Gas nozzles are made by each manufacturer to fit an individual gun, and they are not generally interchangeable. Some nozzles have grooves or fins on the outside to help radiate heat away from them.

have aluminum fins on the outside to radiate heat away from the nozzle. Gas nozzles are not used in guns dedicated for use with the FCAW-S process.

During a welding operation, spatter from molten metal gathers on the inside of the nozzle, **Figure 3-23**. Such contaminated nozzles decrease and deflect the gas flow over the molten metal, causing oxidation of the weld metal. To prevent this condition, coat the inside of the nozzle with an antispatter compound or spray. See **Figure 3-24**. Antispatter dip, **Figure 3-25**, is another solution to spatter buildup in the gas nozzle. A brush or tools can be used to remove spatter buildup between spraying applications. See **Figure 3-26** and **Figure 3-27**.

Workpiece Leads and Work Clamps

The importance of proper workpiece leads and clamps cannot be overemphasized in GMAW or FCAW processes. These workpiece clamps and leads are often incorrectly referred to as ground clamps and leads; however, they are part of the electrical circuit from the negative to positive terminals and do not ground any part of the circuit.

Goodheart-Willcox Publisher

Figure 3-23. A gas nozzle filled with excessive spatter reduces the shielding gas coverage over the weld pool.

Goodheart-Willcox Publisher

Figure 3-25. Use of a special formulated dip reduces the buildup of spatter in the nozzle.

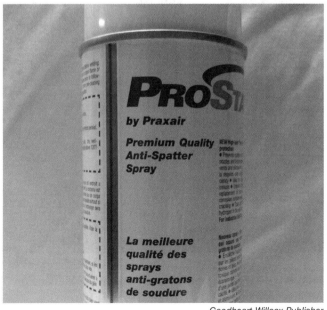

Goodheart-Willcox Publisher

Figure 3-24. Regular use of an antispatter spray reduces nozzle buildup.

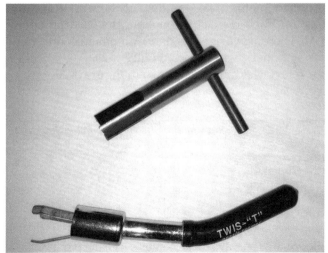

Goodheart-Willcox Publisher

Figure 3-26. Various tools are available to remove the spatter from the gas nozzle.

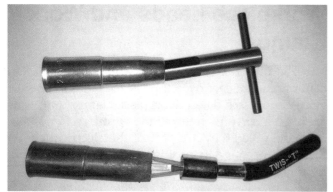

Goodheart-Willcox Publisher

Figure 3-27. Tools that are inserted into a gas nozzle to remove spatter buildup.

The cable must be adequately sized to carry the welding current without overheating. The cable sizes shown in **Figure 3-28** are the minimum sizes—larger cables can always be used. The work clamp must fit tightly to the workpiece or an arc is made when the current flows through the connection. Arc outages may occur where good contact is not made, and the weld joint at these points may be unsatisfactory. Inspect cables regularly for cuts or damage from excessive heating. Cables that are cracked or broken must be replaced.

Connections between the clamp and the cable should be tight, since loose connections do not allow proper current flow. This is a major cause of faulty welding with the GMAW short circuiting mode of welding. To check the system, hold your hand near the cable and connections. If the cable or connections are warm or hot, they require servicing or replacement.

When circular parts mounted on positioners are being welded or the parts are rotated, a rotary work connection eliminates tangling of the work cable and ensures good contact throughout the welding operation. There are various types of clamps and connections for stationary weldments, **Figure 3-29.** A connection for rotary weldments is shown in **Figure 3-30.**

Goodheart-Willcox Publisher

Figure 3-29. Various types of clamps.

Goodheart-Willcox Publisher

Figure 3-30. Rotary work connections are often used on weld positioners.

	Welding Current (amps)	Cable Size (No.)
Recommended Minimum Cable Sizes	100	4
	150	2
	200	2
	250–300	1/0
	300–450	2/0
	500	3/0
	600	4/0

Goodheart-Willcox Publisher

Figure 3-28. Always replace cables with the proper size.

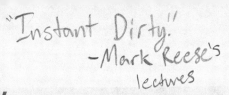
"Instant Dirty"
—Mark Reese's lectures

Summary

- GMAW and FCAW processes use a constant voltage power source.
- Duty cycle is the percentage of time in a ten-minute period that a power source can run at its rated output without overheating and causing damage to the machine.
- NEMA classifies welding power sources into three levels based on the machine duty cycle.
- Controls found on a typical power source are the OCV control and the inductance control.
- Open circuit voltage is the voltage output of the welding power source when no arc is present.
- The separation of the molten drops of metal from the electrode is controlled by the inductance. Squeezing forces are exerted on the electrode due to the current flowing through it.
- A power source should be installed in an area free of dust, dirt, fumes, and moisture.
- The three basic types of wire feeding systems used for GMAW/FCAW are the push type, push-pull type, and spool gun.
- Wire feeder drive rolls are available in various designs based on the size and type of wire used in the welding process.
- Wire wipers are used to remove contaminants from the wire before they enter the drive roll system and the gun.
- The welding gun and gun cables carry the electricity, shielding gas, and electrode wire from the machine to the weld. Workpiece clamps and cables complete the electrical circuit required to create the weld pool.
- The standard GMAW/FCAW welding gun includes the handle and trigger, gas diffuser, contact tip, gas nozzle, and insulator.

Review Questions

Answer the following questions using the information provided in this chapter.

1. Welding power sources are made to operate on _____- or _____-phase primary current.
2. Duty cycle is the percentage of time in a(n) _____ period that a machine can run at its rated output without overheating.
3. In what NEMA class is a 150 A welding machine with a 30% duty cycle?
4. What does OCV stand for?
5. _____ voltage is the voltage output of the welding power source when an arc is present.
6. The separation of molten drops of metal from an electrode is controlled by _____ forces exerted on the electrode due to the current flowing through it.
7. Why is some inductance beneficial in GMAW spray transfer mode?
8. The function of a CV power source is to maintain a preset _____ condition.
9. A power source should be installed in an area free of dust, dirt, fumes, and _____.
10. Terminal blocks should be cleaned with a(n) _____ brush or an abrasive pad.
11. What are the three basic types of wire feeders used for GMAW/FCAW?
12. On an auxiliary wire feeder, the _____ control prevents the wire from sticking in the weld pool.
13. The system that drives the electrode wire from the spool to the gun is either a two-roll or a(n) _____-roll system.
14. Why are wire guides adjusted as close to the rollers as possible?
15. What functions do gun cables perform?
16. *True or False?* Water-cooled guns are used with a large diameter electrode and have a high duty cycle at a high amperage rating.
17. When must contact tips be replaced?
18. *True or False?* Gas nozzles are generally interchangeable.
19. Welding leads must be of a sufficient size to adequately carry the welding current without _____.
20. Loose connections between the clamp and the cable do not allow proper _____ flow.

This welder is using the GMAW process to make a welding repair on the inside of a ship.

Chapter

Shielding Gases and Regulation Equipment

Learning Objectives

After studying this chapter, you will be able to:
- ❏ Explain the functions of shielding gases in GMAW and FCAW.
- ❏ Compare characteristics of welds made with different gases or gas combinations.
- ❏ Identify the gases and gas combinations used on ferrous and nonferrous metals.
- ❏ Describe how shielding gases are supplied and distributed in a GMAW system.
- ❏ Explain how gas flow is regulated.

Technical Terms

argon
carbon dioxide
demurrage
flowmeter
helium
hydrogen
inert gas
manifolds
nitrogen
oxygen
purging
reactive gas
regulator
regulator/flowmeter
shielding gas

Shielding Gases

A *shielding gas* may be a pure gas or a mixture of several gases. Shielding gases perform the following functions in GMAW and FCAW:
- Shield the electrode and the molten metal from the atmosphere.
- Transfer heat from the electrode to the base metal.
- Stabilize the arc pattern.
- Aid in controlling bead contour and penetration.
- Assist in metal transfer from the electrode.
- Assist in the cleaning action of the joint and provide a wetting action.

Shielding gases commonly used with GMAW include argon, helium, carbon dioxide, and oxygen. Argon and helium are *inert gases* (chemically inactive gases that do not combine with any product of the weld area). Inert gases are pure, colorless, and tasteless. They are used as a single gas or a part of a mixed gas combination. Carbon dioxide is not inert but can be used alone or as part of a mixed gas combination. Oxygen is always combined with other gases. Carbon dioxide and argon are the only shielding gases used in FCAW.

Argon

Argon (chemical symbol Ar) is the inert shielding gas most commonly used for GMAW. It is separated from the atmosphere during the production of oxygen and is therefore readily available at low cost. Pure argon is used for GMAW of some nonferrous metals. However, pure argon is not used for FCAW. For FCAW, it is always used in combination with carbon dioxide.

Argon produces narrow bead widths because the arc is more concentrated than that produced in any other gas. The weld penetration is deep in the center of the weld. Argon gas is heavier than air and tends to form a blanket around the electrode and molten metal, reducing spatter and contamination.

Helium

Helium (chemical symbol He) is an inert gas found in natural gas wells and is higher in cost than argon. It is not often used as a single gas in GMAW because of poor metal transfer from the electrode to the weld pool. Helium is commonly used as part of the gas mixture for stainless steel and aluminum applications. Although helium has a higher thermal conductivity than argon, penetration is wider and not as deep. Since helium is lighter than air, the gas tends to rise from the weld area quite rapidly. This makes it necessary to use a higher flow rate to maintain proper shielding.

Carbon Dioxide

Carbon dioxide (chemical symbol CO_2) is a compound *reactive gas* made up of carbon monoxide and oxygen. Reactive gases combine chemically with the weld pool to produce a desirable effect. These individual gases are mixed and then compressed and stored in liquid form, **Figure 4-1**. Although it is not an inert gas, it can be used as a single gas or part of a gas mixture for both GMAW and FCAW. Because CO_2 dissociates (breaks down) at welding temperatures, the arc is more erratic and harsh and tends to have more spatter when the gas is used alone. Regardless of whether it is used as a single or mixed gas, the CO_2 must be welding grade gas. Welding grade gas is dried to remove moisture. Welds made with other grades of carbon dioxide may exhibit various defects.

Oxygen

Oxygen (chemical symbol O_2) is a reactive gas acquired from the atmosphere. It is not used as a single gas for GMAW welding but as part of a gas mixture to attain specific arc patterns. Because the amount of oxygen is a very small percentage of the total gas mixture, the user should purchase the mixture from a supplier. See **Figure 4-2**.

Hydrogen

Hydrogen (chemical symbol H_2) is a reactive gas used in small percentages (1%–5%) when added to argon for shielding stainless steel and nickel alloys. The high thermal conductivity of hydrogen helps produce a fluid weld pool and improved wetting action. This characteristic also permits the use of faster travel speeds. See **Figure 4-3**.

Goodheart-Willcox Publisher

Figure 4-1. A Dewar flask containing liquefied carbon dioxide.

Goodheart-Willcox Publisher

Figure 4-2. A compressed oxygen gas cylinder.

Figure 4-3. A compressed hydrogen cylinder.

Gases and Mixes for GMAW on Ferrous Metals

The following gases and gas mixtures are used for welding ferrous metals:

- 100% carbon dioxide can be used for short circuit gas metal arc welding of steel.
- 99% argon with 1% oxygen should be used only for welding stainless steel in the spray transfer mode in the flat and horizontal positions. Adding oxygen to argon stabilizes the arc while improving the appearance of the weld bead. Groove welding is limited to the flat position. Fillet welds can be done in either flat or horizontal position.
- 98% argon with 2% oxygen can be used in the spray transfer mode for welding carbon steels and stainless steels. Groove welding with spray transfer is limited to the flat position. Fillet welds can be done in either flat or horizontal position.
- 95% argon with 5% oxygen can be used for spray transfer welding of carbon and stainless steels. Groove welding is limited to the flat position. Fillet welds can be done in the flat or horizontal position.
- 90% argon with 10% carbon dioxide and 92% argon with 8% carbon dioxide are the two most common mixtures for pulsed spray transfer welding of carbon steel. These mixtures are sometimes used for applications requiring minimal spatter.

- 75% argon with 25% carbon dioxide is the most common mixture for short circuit gas metal arc welding of carbon steels in all positions. Welds made with this gas mixture have minimum spatter and good penetration features.
- 50% argon with 50% carbon dioxide offers many of the qualities of the 75% argon/25% carbon dioxide mixture but at a lower cost. This is because a smaller amount of the more expensive argon gas is used. When spatter and a deeper penetration can be tolerated, this gas mixture can be used.

Gases and Mixes for GMAW on Nonferrous Metals

The following gases and gas mixtures are used for welding nonferrous metals:

- Argon is used for short circuiting transfer, spray transfer, and pulsed spray transfer welding in all positions on aluminum, nickel-based alloys, and reactive metals. It can also be used for welding some thin-gauge materials in the short circuiting transfer mode.
- Helium gas alone has only limited use in GMAW. Its basic application is in machine welding of heavy aluminum at high welding currents.
- 75% argon with 25% helium is normally used on heavier materials with the spray transfer mode. Penetration is deeper than with pure argon and the weld bead appearance is good.
- 75% helium with 25% argon is used with spray transfer welding of heavier materials. This percentage of helium increases the heat input, which reduces internal porosity. It also provides good wetting of the weld into the base metal.

Special Gas Mixtures

Special gas mixtures include the following:

- 90% helium with 7 1/2% argon and 2 1/2% carbon dioxide is a mixture developed for short circuit gas metal arc welding of stainless steels. Welding can be done in all positions. The argon addition provides good arc stability and penetration. The high helium content provides heat input to overcome the sluggish nature of the weld pool in stainless steel.
- 90% argon with 8% carbon dioxide and 2% oxygen is a mixture used for short circuit gas metal arc welding and pulsed spray welding on carbon steel applications to reduce spatter.

- 60% helium with 35% argon and 5% carbon dioxide is a mixture for short circuit gas metal arc welding of high-strength steel in all positions.

Gases and Gas Mixes for FCAW-G

The electrode manufacturer determines the proper gas or gas mixture for FCAW-G. Manufacturers add certain elements to the flux to obtain mechanical values, produce specific element percentages, deoxidize weld metal, and in some cases, create a weld with low hydrogen content. To ensure that elements transfer across the arc and maintain these additional characteristics, the prescribed gas must be used.

Carbon dioxide is the most popular gas for FCAW when spatter is not a concern on the completed weldment. An argon-carbon dioxide mixture with 50% to 75% argon content is common, with the 75% argon and 25% CO_2 mixture commonly used when switching between FCAW and GMAW on carbon steel applications.

Specialty Gas Mixtures

Some specialty gas mixtures (such as *Stargon*™, *Helistar*®, and *Mig Mix Gold*™) have been developed by gas suppliers for standard and specialized applications. These mixes often allow a broader range of welding parameters than the standard mixes and thus reduce welding costs. See **Figure 4-4**.

Goodheart-Willcox Publisher

Figure 4-4. A proprietary tri-mix blend of helium, argon, and carbon dioxide, called *Helistar*®.

Selecting the Proper Gas Mixture

As an essential variable in the welding process, the selection of the proper shielding gas or mixture depends on all the other variables involved. When selecting a gas or gas mixture, consider the following factors:

- Mode of metal transfer.
- Base metal type.
- Base metal thickness.
- Joint design.
- Weld position.
- Filler material composition.
- Filler material size.
- Chemical composition of the desired weld metal.
- Weld metal quality.

GMAW shielding gases and their applications are listed in **Figure 4-5A**. The gases for short-circuit GMAW are listed in **Figure 4-5B**. **Figure 4-5C** lists the selection of gases for GMAW with spray transfer.

Purge Gas Applications

Purging on the root side of the weld may be desired or necessary to protect it from atmospheric contamination. A purge gas is used to displace the atmosphere from the adjacent weld zone under the weld opening or inside a pipe or tube. Argon, **nitrogen,** or helium is used for this purpose. Where the application requires a high level of protection, always use argon or helium. In many cases, however, nitrogen can be used as an inexpensive initial purge gas before using the more expensive gases. The nitrogen displaces a large percentage of atmospheric oxygen at a reduced cost. Nitrogen gas is available in either a compressed gas or liquid state, **Figure 4-6**. The purge gas may also be used to protect the back of fillet or groove welds that do not require full penetration of the root side of the weld. See **Figure 4-7**.

Gas Purity

Inert shielding gases for welding applications are refined to high purity specifications. Cylinder argon has a minimum purity of 99.996% and contains a maximum of about 15 parts per million (ppm) moisture (a dew point temperature of –73°F maximum). Driox® brand argon has a minimum purity of 99.998% and a moisture content of less than six ppm. Helium is produced to a minimum purity of 99.995%. Generally, helium contains less than 15 ppm of moisture. At these levels of refinement, impurities usually cannot be detected during welding.

Shielding Gas	Chemical Behavior	Typical Application
Argon	Inert	Virtually all metals except steels.
Helium	Inert	Aluminum, magnesium, and copper alloys for greater heat input and to minimize porosity.
Ar + 20–80% He	Inert	Aluminum, magnesium, and copper alloys for greater heat input and to minimize porosity (better arc action than 100% helium).
Nitrogen		Greater heat input on copper (Europe).
Ar + 25–30% N_2		Greater heat input on copper (Europe); better arc action than 100% nitrogen.
Ar + 1–2% O_2	Slightly oxidizing	Stainless and alloy steels; some deoxidized copper alloys.
Ar + 3–5% O_2	Oxidizing	Carbon and some low-alloy steels.
CO_2	Oxidizing	Carbon and some low-alloy steels.
Ar + 20–50% CO_2	Oxidizing	Various steels, chiefly short circuiting mode.
Ar + 10% CO_2 + 5% O_2	Oxidizing	Various steels (Europe).
CO_2 + 20% O_2	Oxidizing	Various steels (Japan).
90% He + 7.5% Ar + 2.5% CO_2	Slightly oxidizing	Stainless steels for good corrosion resistance, short circuiting mode.
60 – 70% He + 25 – 35% Ar + 4 – 5% CO_2	Oxidizing	Low-alloy steels for toughness, short circuiting mode.

A

Metal	Shielding Gas	Advantages
Carbon steel	75% argon + 25% CO_2	Less than 1/8 in. (3.2 mm) thick; high welding speeds without burn-thru; minimum distortion and spatter.
	75% argon + 25% CO_2	More than 1/8 in. (3.2 mm) thick; minimum spatter; clean weld appearance; good puddle control in vertical and overhead positions.
	CO_2	Deeper penetration; faster welding speeds.
Stainless steel	90% helium + 7.5% argon + 2.5% CO_2	No effect on corrosion resistance; small heat-affected zone; no undercutting; minimum distortion.
Low-alloy steel	60–70% helium + 25–35% argon + 4–5% CO_2	Minimum reactivity; excellent toughness; excellent arc stability, wetting characteristics, and bead contour; little spatter.
	75% argon + 25% CO_2	Fair toughness; excellent arc stability, wetting characteristics, and bead contour; little spatter.
Aluminum, copper, magnesium, nickel, and their alloys	Argon & argon + helium	Argon satisfactory on sheet metal; argon-helium preferred on thicker sheet material (over 1/8 in. [3.2 mm]).

B

Goodheart-Willcox Publisher

Figure 4-5. A—GMAW shielding gases and applications. B—Recommended shielding gases for GMAW with short circuiting transfer.

Most steels and copper alloys show high tolerances for various amounts of contaminants. Aluminum and magnesium are sensitive to gas purity and exhibit severe defects if contaminated gas is used. Still other materials, such as reactive metals, have extremely low tolerances for contaminants in the inert gases. High purity standards are maintained by gas suppliers to ensure that shielding gases are more than adequate for the most severe application.

Gas Supply

Shielding gases are supplied in cylinders of various sizes for shop use. See **Figure 4-8** and **Figure 4-9**. Dewar flasks (insulated, pressurized containers for liquefied gas) and high-pressure gas cylinders mounted on trailers are used where large volumes of gases are required.

In most cases, gas is sold through a distributor by the total quantity in cubic feet of each type of gas.

Metal	Shielding Gas	Advantages
Aluminum	Argon	0 to 1 in. (0 to 25 mm) thick; best metal transfer and arc stability; least spatter.
	35% argon + 65% helium	1 to 3 in. (25 to 76 mm) thick; higher heat input than straight argon; improved fusion characteristics with 5XXX series Al-Mg alloys.
	25% argon + 75% helium	Over 3 in. (76 mm) thick; highest heat input; minimizes porosity.
Magnesium	Argon	Excellent cleaning action.
Carbon steel	Argon + 1–5% oxygen	Improves arc stability; produces a more fluid and controllable weld pool; good coalescence and bead contour; minimizes undercutting; permits higher speeds than pure argon.
	Argon + 3–10% CO_2	Good bead shape; minimizes spatter; reduces chance of cold lapping; cannot be used out of position.
Low-alloy steel	Argon + 2% oxygen	Minimizes undercutting; provides good toughness.
Stainless steel	Argon + 1% oxygen	Improves arc stability, produces a more fluid and controllable weld puddle, good coalescence and bead contour; minimizes undercutting on heavier stainless steels.
	Argon + 2% oxygen	Provides better arc stability; coalescence, and welding speed than 1% oxygen mixture for thinner stainless steel materials.
Copper, nickel, and their alloys	Argon	Provides good wetting, decreases fluidity of weld metal for thickness up to 1/8 in. (3.2 mm).
	Argon + helium	Higher heat inputs of 50% & 75% helium mixtures offset high heat dissipation of heavier gases.
Titanium	Argon	Good arc stability; minimum weld contamination; inert gas backing is required to prevent air contamination on back of weld area.

Goodheart-Willcox Publisher

Figure 4-5. C—Recommended shielding gases for GMAW with spray transfer.

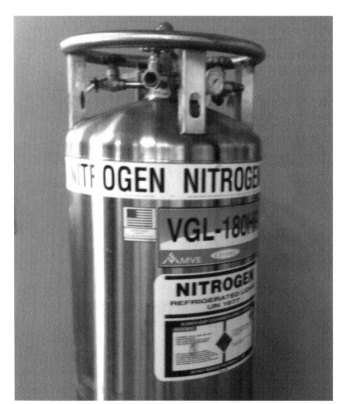

Goodheart-Willcox Publisher

Figure 4-6. A Dewar flask containing liquefied nitrogen for purging applications.

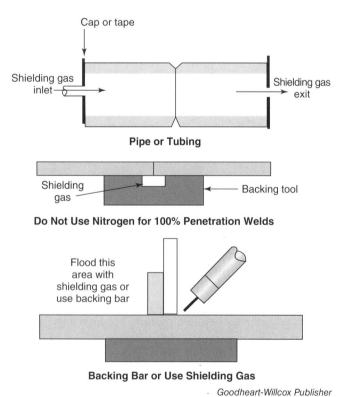

Goodheart-Willcox Publisher

Figure 4-7. In purging applications, nitrogen is used for carbon steels. Argon or helium is used for purging in all aluminum and stainless steel applications.

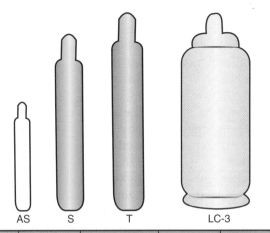

Cylinder Style	Contents Cubic Ft.	Full Pressure of Cylinder at 70°F (21°C)	Height	O.D.
AS	78	2200	35	7 1/8
S	150	2200	51	7 3/8
T	330	2640	60	9 1/4
LC-3	2900	55	58	20

Goodheart-Willcox Publisher

Figure 4-8. Types and capacities of cylinders and Dewar flasks supplied to industry.

Since gas is distributed in cylinders, the distributor also charges a *demurrage* (rental) fee on each cylinder used. When large quantities of gases are needed, the liquefied form is the most economical because fewer cylinders are used. See **Figure 4-10**.

Gas Storage

Storage of shielding gas cylinders and containers should be carefully controlled. Gas cylinders or containers should always be stored outside or in a well-ventilated area. They are often stored with other high-pressure cylinders, such as oxygen. Cylinders should be stored upright (in a vertical position), with the safety cap in place, and secured to a rigid support to keep them from falling over. The type or mixture of the cylinder's gas contents should be properly and clearly identified on the cylinder label. Follow all safety precautions to avoid injury. Remember, inert gases do not contain oxygen and therefore will not support life. You cannot see, smell, or taste inert gases.

Goodheart-Willcox Publisher

Figure 4-9. Various size cylinders ready to be loaded on the distributor's supply truck.

Goodheart-Willcox Publisher

Figure 4-10. A Dewar flask containing liquefied argon may be used connected to a manifold system, with or without a shielding gas mixer station.

Gas Distribution

Gases are distributed to the welding area in several ways. A bank of cylinders can be connected together. The cylinders are located in a convenient place and the gas is piped to the welding area. With this arrangement, one bank may be used until empty, and then another can be placed into use. Empty cylinders, in turn, can be replaced without affecting the bank in use.

Manifolds are often used to distribute gases to the welding area from the supply area. A manifold system allows two or more cylinders to be regulated and connected through a distribution pipeline system and the gas delivered to several welding stations as needed. These systems are usually connected at an outside location for ease of delivery and connection. Through the use of manifolds, the number of individual cylinders required at the welding station is reduced. See **Figure 4-11**.

The distribution system must be leak free to maintain the purity of the gas. Therefore, all of the cylinder fittings must be cleaned before installation and properly seated into the regulators. High-pressure connectors, tubing, and pipe connectors must be protected to prevent entry of foreign materials, water, or oil when the system is not in use. Threaded plastic adapters or tape may be placed over any unused openings for this purpose. Use Teflon® tape when connecting any threaded fittings.

Testing for Gas Leaks

Before a system can be used, it must be tested for leaks at a pressure higher than the normal operating pressure. One testing method is a special leak test solution, **Figure 4-12**, which is applied while the pipe is under pressure. Because the solution is sensitive to gas flow, it forms bubbles wherever there is a leak.

Another testing method is to apply pressure to the system and note the test pressure. The inlet pressure valve is then closed and the pressure gauge is observed closely. A drop in pressure indicates that a leak is present.

Prior to placing the manifold in use, the lines must be purged to remove flux vapors and moisture. Nitrogen gas can be used as the initial purging gas. Connect the gas to the manifold, set the flow rate on the other end of the pipe to 5 cubic feet per hour (cfh) and purge until analyzer tests show the system is clear of oxygen. When the test shows clear, connect the argon to the line and purge at the same rate for several hours before welding. If the system requires repair, disassembly, or modification, it must be purged and tested again prior to use.

Gas Regulation

Regulation of shielding gases is accomplished with several types of special equipment. Gases distributed through a manifold require *regulators*. Regulators reduce the pressure from the cylinder to the desired manifold pressure level, usually in the range of 20 psi to 50 psi. A *flowmeter* is used at the welding station to regulate the flow of gas to the welding gun. When a cylinder is used at the welding station itself, a *regulator/flowmeter* or flow-regulator reduces pressure from the cylinder and regulates the volume of flow to the gun. See **Figure 4-13** and **Figure 4-14**. A regulator/flowmeter or a flowmeter with a ball tube should always

Goodheart-Willcox Publisher

Figure 4-11. This manifold system allows several cylinders to distribute various shielding gases to numerous welding stations.

Goodheart-Willcox Publisher

Figure 4-12. A special solution is used to check for leaks in the gas distribution system.

Goodheart-Willcox Publisher

Figure 4-13. A single cylinder regulator/flowmeter. The regulator gauge measures the cylinder pressure. The attached flowmeter measures the volume of gas to the welding gun in cfh. The volume is read from the top of the ball in the vertical tube and adjusted with the knob on the top.

Goodheart-Willcox Publisher

Figure 4-14. A flow-regulator is an economy version of a flowmeter. The volume is adjusted with the knob on the front of the gauges.

be installed with the ball tube in the vertical position for proper operation. The amount of flow is indicated at the top of the ball unless otherwise indicated.

Regardless of the type of gas supply (cylinder, Dewar flask, manifold), a surge of gas exits from the gas nozzle when the gas flow solenoid valve is opened. This is due to the pressure buildup when the gas is not flowing. This initial gas surge lasts for several seconds until the excess pressure is reduced. To eliminate this condition, a specially designed surge check valve can be used.

Gas Mixing

Welding gases can be mixed at a shielding gas mixing station and stored before they enter the manifold. A mixing station is also used for several robotic applications to achieve desired percentages. The use of an onsite gas shielding gas mixer is helpful where

large volumes of gases are needed. **Figure 4-15** shows an outside manifold system set up for argon and carbon dioxide gas cylinders. The manifold regulates the pressure to the gas mixer. See **Figure 4-16**.

Mixture and Purity Testing

Gas analyzers are used to test for proper mixes at the welding stations. These instruments can also be used to check for leaks and to establish that pipes or vessels have been adequately purged before welding.

Goodheart-Willcox Publisher

Figure 4-15. This gas manifold system is designed for the large volumes of argon and carbon dioxide needed for a gas mixing station.

Goodheart-Willcox Publisher

Figure 4-16. This small mixer is used on single-station installations. Larger mixers are employed in multiple station installations.

Summary

- Shielding gases shield the electrode and the molten metal from the atmosphere, transfer heat from the electrode to the base metal, stabilize the arc pattern, aid in controlling bead contour and penetration, assist in metal transfer from the electrode, aid in the cleaning action of the joint, and provide a wetting action for the weld pool.
- Shielding gases can be used separately or in combination.
- Carbon dioxide and argon are the only shielding gases used in FCAW.
- Inert shielding gases are argon and helium. Reactive shielding gases are carbon dioxide, oxygen, and hydrogen.
- The most common purge gas for carbon steel is nitrogen.
- The proper gas or mixture depends on the type and thickness of the base metal, process, metal transfer mode, joint design and position, electrode type and size, chemical requirements, and quality requirements of the completed weld.
- Shielding gas flow is reduced in pressure from the cylinder with a regulator/flowmeter. The flowmeter controls the volume of gas (in cfh) through the welding gun to the weld zone.
- Welding gases can be mixed at a mixing station and stored before they enter the manifold.

Review Questions

Answer the following questions using the information provided in this chapter.

1. Shielding gases perform several functions, including shielding the electrode and the molten metal from the atmosphere and stabilizing the _____ pattern.
2. List two inert gases used in the GMAW process.
3. List two reactive gases used in the GMAW process.
4. _____ is always used as a part of a gas combination, not as a single gas.
5. What shielding gases are used for FCAW-G?
6. Argon gas is _____ than air and tends to form a blanket around the molten weld pool.
7. _____ gas is lighter than air and requires higher flow rates than argon.
8. Carbon dioxide is a manufactured gas made by combining _____ and _____.
9. Why do welds have more spatter when they are made with carbon dioxide alone as a shielding gas?
10. The most common shielding gas mixture for welding carbon steels with the short circuiting transfer mode is _____.
11. Which shielding gas mixture is normally used to weld heavier materials with the spray transfer mode?
12. _____ can be used as an inexpensive initial purge gas before using argon or helium.
13. List two metals that exhibit severe porosity if contaminated shielding gas is used.
14. Insulated, pressurized containers for liquefied gas are called _____.
15. Where should gas containers be stored?
16. Gas cylinders should be stored in a vertical position with the _____ in place.
17. What is a *manifold*?
18. Manifolds are purged to remove flux vapors and _____.
19. What is the function of a regulator on a manifold?
20. At welding stations, _____ are used to test for proper gas mixes.

GMAW and FCAW Electrodes

Learning Objectives

After studying this chapter, you will be able to:

- ❑ Describe how electrode wire materials are made.
- ❑ Identify the specifications that apply to GMAW and FCAW electrodes.
- ❑ Describe the forms and sizes of filler materials.
- ❑ Explain how manufacturers and welders prevent contamination of electrode wire.
- ❑ Identify factors that determine selection of the proper electrode.

Technical Terms

annealing
cast
helix
hot-drawn
layer wound
level-layer wound

Manufacture of GMAW and FCAW Electrodes

To produce acceptable welds, electrode filler materials used in the GMAW and FCAW processes must be of the highest quality. For this reason, manufacturers of electrode wire employ specialized machines, processes, and inspections. They supply the electrode wire in roll form to fit various types of wire feeders.

Selection of material for the manufacture of GMAW welding wire is based on several factors, including the following:

- • Chemical composition.
- • Mechanical properties.
- • Notch toughness values.
- • Impurity level limits.

Material selected at the primary mill is *hot-drawn* (made into a large coil of heavy wire at the steel mill while hot), then pulled through reducing dies to specified sizes before the mill scale is cleaned off. Further reduction of the wire is then done cold through another series of reducing dies. *Annealing* (softening of the metal by heating and slow cooling) and cleaning operations are performed as the wire is further reduced in size. If the wire is not properly annealed, it becomes brittle and subject to breakage during the coiling process. Stainless steel electrode wire may need to be annealed several times during the process due to the work hardening effect of the reducing dies.

Lubricants are applied to the dies during the drawing process to reduce die wear and decrease friction between the wire and the dies. Lubricants also carry away heat produced in the dies during the drawing operation.

In FCAW, a tubular electrode and an internal flux are used to deposit the weld metal. The flux mixture consists of deoxidizers, slag formers, and arc stabilizers. The electrode base material is formed into a *U* shape, flux is added, and a tube is formed. The tube is drawn (reduced) to a smaller size, compressing the flux into a tight matrix. See **Figure 5-1**.

The chemical composition of the electrode and flux can be tightly controlled to yield an exact composition in the weld. The flux is sealed in the electrode tube, extending storage life and reducing handling problems.

During the drawing process of GMAW solid wire electrodes, several types of defects can occur that affect the quality of filler materials. See **Figure 5-2**. These defects include the following:

- Cracking.
- Seaming.
- Center bursting.
- Oxide formation.
- Overlapping.

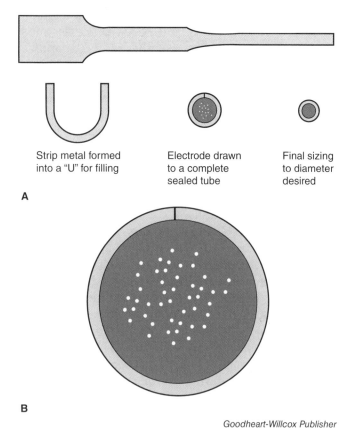

Strip metal formed into a "U" for filling

Electrode drawn to a complete sealed tube

Final sizing to diameter desired

A

B

Goodheart-Willcox Publisher

Figure 5-1. Manufacture of a tubular electrode. A—Metal is formed into a *U*, filled with core material, drawn into a tube, and sized to a precise diameter. B—A cross section of core material consisting of powdered metals, vapor-forming material, deoxidizers, scavengers, and slag formers.

Filler materials that exhibit any of these defects should not be used. After drawing, another process cleans the wire of all surface impurities. The filler material is then processed for shipment to the user.

Defects commonly found on the surface of a solid electrode are not evident on a tubular FCAW electrode. Because the tube is drawn from thin sheet metal, the reduction area is much less than when a solid material is drawn. After the tube is formed, it is heat-treated (if required), cleaned, and packaged for final use.

Electrode Specifications

The AWS issues specifications for GMAW solid electrodes and FCAW tubular electrodes used with carbon steels, low-alloy steels, corrosion-resisting chromium, and chromium-nickel steels. Other electrodes made for surfacing, cladding, and repairing cast iron are made to a commercial or AWS specification. AWS specifications for steel and stainless steel electrodes are as follows:

- AWS A5.18 *Carbon Steel Electrodes for Gas Shielded Arc Welding.*
- AWS A5.20 *Carbon Steel Electrodes for Flux Cored Arc Welding.*
- AWS A5.22 *Stainless Steel Flux Cored and Metal Cored Welding Electrodes and Rods.*
- AWS A5.28 *Low-Alloy Steel Electrodes and Rods for Gas Shielded Arc Welding.*
- AWS A5.29 *Low-Alloy Steel Electrodes for Flux Cored Arc Welding.*

A number following the specification number indicates the year of publication. Always refer to the

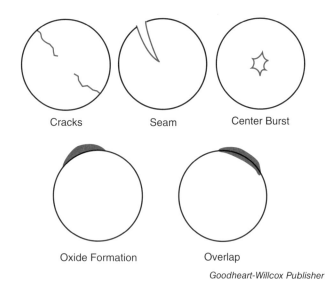

Cracks   Seam   Center Burst

Oxide Formation   Overlap

Goodheart-Willcox Publisher

Figure 5-2. Typical defects that occur in welding wire electrodes during manufacturing.

most recent specifications available. Carbon steel and low-alloy steel electrodes for GMAW and FCAW are identified by an AWS coded system. See **Figure 5-3** and **Figure 5-4**. Chromium and chromium-nickel steels use a base metal system for identification, **Figure 5-5**. Additional information for electrodes is provided in the *Reference Section.*

Many electrodes are manufactured for specific purposes, such as joining, cladding, hardfacing, or building up welds. They are made to company specifications or special specifications that govern the chemical content, mechanical properties, and usability of the filler material.

To ensure that all electrodes meet equal standards of quality, AWS specifications establish requirements for scope, classification, acceptance, manufacture, and testing. Additional requirements may include chemical tests, impact test results, and special packaging.

Electrode manufacturers guarantee only that their products meet AWS specifications. They replace defective electrodes but do not guarantee acceptable results because they cannot control the welding process.

Electrode manufacturers make welding filler material for three major areas:

- **General use.** This wire meets specification requirements. However, no record of chemical composition, strength level, or other requirements is submitted to the user when the wire is purchased.
- **Rigid control use.** These electrodes are used in fabrication processes that require rigid control over the filler material. Wire used under these conditions may require a Certificate of Conformance with the purchase of the material. This certificate is a statement that the filler material meets all of the requirements of the material specification. All of the stock is

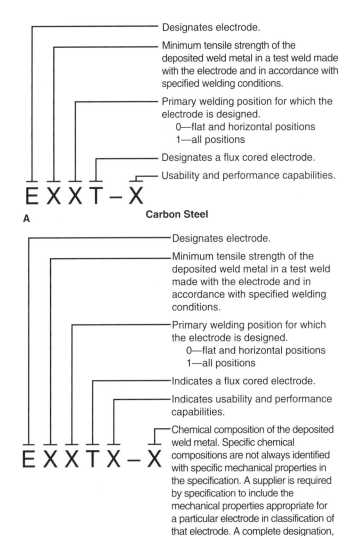

A **Carbon Steel**

- Designates electrode.
- Minimum tensile strength of the deposited weld metal in a test weld made with the electrode and in accordance with specified welding conditions.
- Primary welding position for which the electrode is designed.
  0—flat and horizontal positions
  1—all positions
- Designates a flux cored electrode.
- Usability and performance capabilities.

E X X T – X

B **Low-Alloy Steel**

- Designates electrode.
- Minimum tensile strength of the deposited weld metal in a test weld made with the electrode and in accordance with specified welding conditions.
- Primary welding position for which the electrode is designed.
  0—flat and horizontal positions
  1—all positions
- Indicates a flux cored electrode.
- Indicates usability and performance capabilities.
- Chemical composition of the deposited weld metal. Specific chemical compositions are not always identified with specific mechanical properties in the specification. A supplier is required by specification to include the mechanical properties appropriate for a particular electrode in classification of that electrode. A complete designation, for example, is E80T5 Ni$_3$. EXXT5 Ni$_3$ is not complete.

E X X T X – X

Goodheart-Willcox Publisher

Figure 5-4. AWS classification of FCAW electrodes. A—Carbon steel. B—Low-alloy steel.

E R XX S – X

- Designates electrode.
- Designates rod, meaning the electrode can also be used for a filler rod for GTAW applications.
- Designates minimum tensile strength in thousands of pounds per square inch of the deposited weld metal in a test weld made with the electrode and in accordance with specified welding conditions.
- Designates a solid electrode wire.
- Designates the specific alloy and deoxidizer(s) that make up the carbon steel electrode's chemical composition.

Goodheart-Willcox Publisher

Figure 5-3. AWS classification of GMAW electrodes for carbon steel.

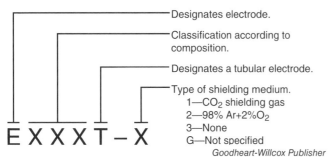

Designates electrode.

Classification according to composition.

Designates a tubular electrode.

Type of shielding medium.
1—CO$_2$ shielding gas
2—98% Ar+2%O$_2$
3—None
G—Not specified

E X X X T – X

Goodheart-Willcox Publisher

Figure 5-5. AWS classification of FCAW chromium and chromium–nickel steel electrodes.

identified by heat numbers, lot numbers, or code numbers located on the wire roll package. On some work, the buyer may require that these numbers be recorded wherever the material is used in a welding application.

- **Critical use.** Some welding operations, such as those involving aircraft, nuclear reactors, and pressure vessels, require very close control of the filler material's chemistry. A Certified Chemical Analysis report is the actual chemical analysis of the individual heat or lot of material. Many electrode manufacturers currently use specific trade names and number suffixes to distinguish critical use wire from the general use electrodes. Records are maintained during fabrication cycles wherever the specific materials are used. If a joint fails due to either the filler material or the base metal, other welded joints in the weldment or system can be evaluated for possible replacement.

Electrode Material Form

Electrode wire used in GMAW and FCAW are wound on spools or inside large drums, depending on the application. Standard spools are 4″, 8″, or 12″ in diameter and are made of plastic, wood, or formed metal wire. The spools are disposable after use. See **Figure 5-6**. For extended use and robotic

applications, the electrode wire is packaged in large drums, **Figure 5-7**.

The wire is either *level-layer wound* or *layer wound*. Level-layer wound electrode wire, **Figure 5-8**, is used for all soft and cored wire. The electrode is carefully spooled onto the reel with each circle next to the previous one and each layer directly over the

Goodheart-Willcox Publisher

Figure 5-7. Large quantities of electrode wire are coiled inside boxes or drums for extensive use or robotic applications.

Goodheart-Willcox Publisher

Figure 5-8. Soft electrode wires like aluminum and cored wire are level-layer wound on the spool.

Goodheart-Willcox Publisher

Figure 5-6. Electrode wire is available on several different size spools to fit most application needs.

one beneath. Layer wound electrode wire, which is used for GMAW hard wires, has one layer atop the other without any distinct pattern. See **Figure 5-9**. The type of winding depends on the material type and the application. See **Figure 5-10**.

Welding wires are furnished in standard sizes, which include the following:

- 0.025″ diameter.
- 0.030″ diameter.
- 0.035″ diameter.
- 0.045″ diameter.
- 3/64″ diameter (soft wires).
- 0.062″ diameter.

Different types and sizes of wire have varying quantities of wire on the spool. Nonferrous materials always have less metal per spool than ferrous materials.

To prevent wire-feeding problems during welding, the wire must meet cast and helix requirements. The terms *cast* and *helix* define the characteristics of any form of continuous wire as it comes from the spool or coil. See **Figure 5-11**. *Cast* is the diameter of one complete circle of wire from the spool as it lies on a flat surface. With hard wire, this diameter tends to be larger than the diameter on the spool. *Helix* is the maximum height of any point of this circle of wire above the flat surface. These two dimensions are critical to the proper feeding of filler material. Improper dimensions affect the entire wire feed system and can cause the tangled condition called *birdnesting*. Other adverse effects of improper wire dimensions are arc outages, severe liner wear, and contact tip wear.

Identification and Packaging

Spools or coils of wire are identified by labels or tapes attached to the inside of the spool hub or to the flange. See **Figure 5-12**.

Goodheart-Willcox Publisher

Figure 5-9. Hard electrode wire like steel and stainless steel are layer wound on the spool.

Used with permission of The Lincoln Electric Company, Cleveland, Ohio, U.S.A.

Figure 5-10. Various size rolls of metal cored electrode wire.

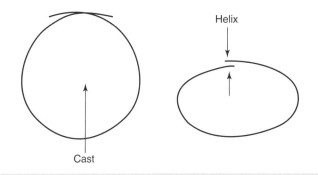

		Cast				Helix	
Spool Size	Wire Types	Min.		Max.		Max.	
		in	mm	in	mm	in	mm
4″ (100 mm)	Low-alloy, stainless and nickel alloy	6*	150	9	230	½	13
	Aluminum	†		6	150	1	25
8″ (200 mm)	Low-alloy, stainless and nickel alloy	15*	380	30	760	1	25
	Low-alloy, stainless and nickel alloy	15*	380	30	760	1	25
12″ (300 mm)	Aluminum	†		15	380	1	25

Standards for Cast and Helix

* Measured on outside strand of full spool
† Diameter of wire level from which sample is taken

Goodheart-Willcox Publisher

Figure 5-11. Cast and helix dimensions and tolerances. Cast is the diameter of one complete circle of wire as it lies on a flat surface. Helix is the maximum height of any point of this circle of wire above the flat surface.

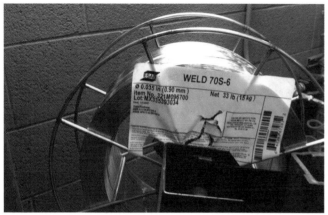

Figure 5-12. Spools or coils of wire are identified by labels or tapes attached to the inside of the spool hub or to the flange.

Manufacturers of electrode wire use a variety of packaging methods to protect the material during shipment and storage. The spools are wrapped in a special paper, foil, or in plastic to protect the wire. See **Figure 5-13** and **Figure 5-14**. The spools are then packed in cardboard cartons or metal cans.

Electrode Use

Oil, moisture, grease, soot, and salts deposited from improper handling easily contaminate filler materials. Dirty gloves, rags, or work surfaces readily contaminate the welding wire whenever they contact it. Such contamination often causes defects in welds. Reworking a rejected weld is costly.

Figure 5-13. This metal cored electrode wire is sealed in an aluminum foil wrap to prevent oxidation.

Figure 5-14. This spool of stainless steel electrode wire is sealed in a plastic bag and packed into a cardboard box.

Electrode material packaging is designed to prevent contamination of the material during shipment. Preventing contamination after the package is opened is the responsibility of the user. The simplest way to avoid contamination is to keep the filler material clean. Follow these practices:

- Keep the material packaged as long as possible.
- Open packages only when needed.
- Store all unsealed electrode wire in a heated cabinet.
- Handle the material as little as possible, and then only with clean gloves.
- Work in a clean and dry area.
- Remove wire spools from machines whenever the machines are to be idle for an extended period of time.
- When changing electrode spools, inspect drive rollers, guide tubes, liners, and contact tips for wear and need for replacement. Remove any metal particles to avoid contamination of the filler material.

When high-quality welds are required, contaminated electrode wire cannot be tolerated. Requiring the electrode manufacturer to clean, inspect, and package the material according to specifications does little good if the material is contaminated by improper handling prior to use. When a roll of steel wire is left exposed to the elements, the wire is no longer usable and must be discarded. See **Figure 5-15**. Once it is installed on the wire feeder, keep the electrode wire covered when not in use. The box that the wire came in makes an ideal cover after one end is removed, **Figure 5-16**.

Wire must be properly handled before being stored. See **Figure 5-17**. To prevent overlapping and tangling of the wire, carefully tie the end of the wire to the spool after removing the wire from the feeder guides. See **Figure 5-18**.

Goodheart-Willcox Publisher

Figure 5-15. This spool of wire has been left exposed. It has oxidized and must be discarded.

Goodheart-Willcox Publisher

Figure 5-16. The cardboard box that the wire is shipped in makes a good cover when installed on an open wire feeder while it is not in use.

Electrode Selection

Selecting the correct welding wire is important in the GMAW process. The chosen wire, in combination with the shielding gas, produces the deposit chemistry that determines the final physical and mechanical properties of the weld. Selection of a welding wire for a particular welding operation is described in the chapters relating to welding specific base materials. The following basic factors influence the choice of welding wire:

- Base metal chemical composition.
- Base metal mechanical properties.
- Type of shielding gas (if required).
- Type of weld joint design.
- Position of the weldment.
- Service use of the weldment.

Goodheart-Willcox Publisher

Figure 5-17. This roll of wire was removed from the machine, and the loose end of the wire was released before being secured on the reel. When the wire retightened on the roll, it overlapped and tangled to the point where it had to be discarded.

Goodheart-Willcox Publisher

Figure 5-18. Carefully tie off the end of the electrode wire when removing the spool from the wire feeder.

Summary

- Manufacturers produce welding electrode wire for the GMAW and FCAW processes under strict quality guidelines for general, rigid control, and critical use applications.
- GMAW electrodes are drawn to a solid wire size. FCAW electrodes are tubular with a flux inside the hollow center.
- GMAW and FCAW electrodes are supplied as a continuous coil of wire in various sizes and diameter spools.
- Hard GMAW electrode wires are furnished in a layer wound spool. Soft GMAW and FCAW electrodes are supplied in a level-layer wound spool to prevent the deformation of the electrode.
- The type of wire and size of the storage spool determine the proper cast and helix of the electrode wire.
- Electrode wires should be properly handled and stored to prevent contamination to the weld.
- Selection of the proper electrode wire is determined by the base metal chemical composition and mechanical properties, the use of and type of shielding gas, the type of weld, position of the weld, and the service use of the weldment.

Review Questions

Answer the following questions using the information provided in this chapter.

1. What four factors determine the selection of material for the manufacture of GMAW welding wire?
2. As wire is drawn through a die to reduce its diameter, it becomes hard and must be _____ to make it soft.
3. List five types of defects produced during the drawing operation that can make wire unusable.
4. The AWS specification for stainless steel electrodes for GMAW and FCAW is _____.
5. What is a Certificate of Conformance?
6. *True or False?* Level-layer wound electrode wire is used for GMAW hard wires.
7. Improper _____ and _____ dimensions can cause wire-feeding problems, such as birdnesting.
8. What effect does filler material contamination have on welds?
9. Where should unsealed filler materials be stored?
10. List six factors influencing the selection of the proper electrode.

Chapter

Weld Joints, Weld Types, and Welding Symbols

Objectives

After studying this chapter, you will be able to:

❑ Identify the basic types of weld joints.

❑ Recognize the types of welds made for each type of joint.

❑ Describe common weldment configuration designs.

❑ Understand common welding terminology and symbols.

❑ List factors involved with joint design.

❑ Describe how weld designs and joints are evaluated.

Technical Terms

arrow side	joint
AWS welding symbol	other side
backing bar	plug weld
buttering	reference line
cladding	stringer beads
double weld	surfacing
fillet weld	undercut
groove weld	weld symbols
interference fit	

Weld Joints

Two major weld profiles are used in most welding applications—the fillet weld and the groove weld. A *fillet weld* is approximately a triangular cross section used to connect perpendicular or uneven joint surfaces at right angles to each other in a lap joint, T-joint, or corner joint. A *groove weld* is used to connect two joint surfaces or edges against each other. See **Figure 6-1**. Along with the single weld bead, these profiles are used in all joint types and positions. The American Welding Society defines a *joint* as "the manner in which materials fit together." As shown in **Figure 6-2**, there are five basic types of weld joints:

- Butt joint.
- T-joint.
- Lap joint.
- Corner joint.
- Edge joint.

Joint Type	Groove Weld	Fillet Weld
Butt	X	
T		X
Lap		X
Corner	Outside	Inside
Edge	X	

Goodheart-Willcox Publisher

Figure 6-1. Most weld joints fit into two weld profiles, the fillet weld, the groove weld, or a combination of the two.

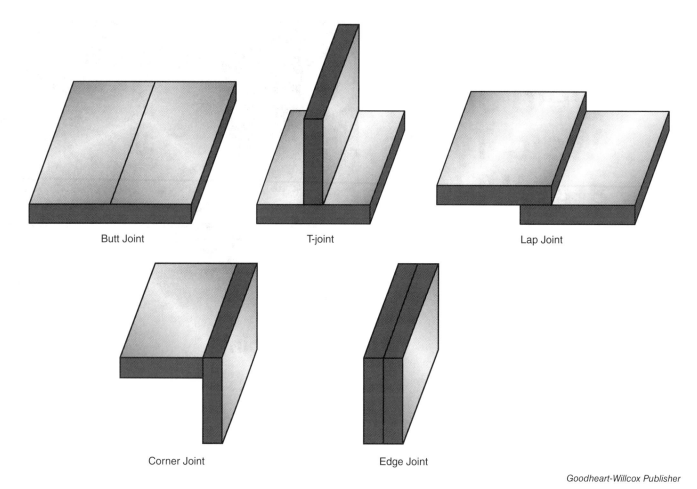

Butt Joint   T-joint   Lap Joint

Corner Joint   Edge Joint

Goodheart-Willcox Publisher

Figure 6-2. Five basic types of weld joints used in FCAW.

Plug and slot welds are often used in lap weld applications to add strength to the weldments where the material is too thick for spot welds. A *plug weld* is a weld made in a circular hole in one member of a joint fusing that member to another member. A slot weld is a weld made in an elongated hole in one member of a joint fusing that member to another member. The hole may be open at one end. Plug and slot welds are also used where additional welding is required to reinforce the primary joint. See **Figure 6-3**.

Weld Joint Preparation

Initial preparation of weld joints can be done in a number of ways:
- Shearing.
- Casting.
- Forging.
- Machining.
- Stamping.
- Filing.
- Routing.
- Oxyacetylene cutting (thermal cutting process).
- Plasma-arc cutting (thermal cutting process).
- Grinding.

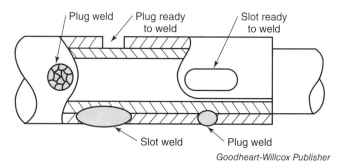

Plug weld   Plug ready   Slot ready
              to weld   to weld

Slot weld   Plug weld

Goodheart-Willcox Publisher

Figure 6-3. Plug and spot welds can be made in many combinations.

Final joint preparation before welding is discussed in subsequent chapters that cover welding of particular metals.

Weld Types

Various types of welds can be made in each of the basic joints. They include the following:

Butt joint. (Figure 6-4)
- Square-groove butt weld.
- Bevel-groove butt weld.

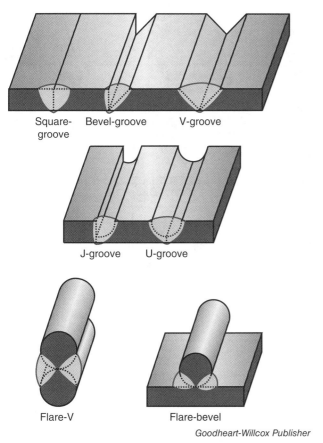

Goodheart-Willcox Publisher

Figure 6-4. Types of welds that can be made in a basic butt joint.

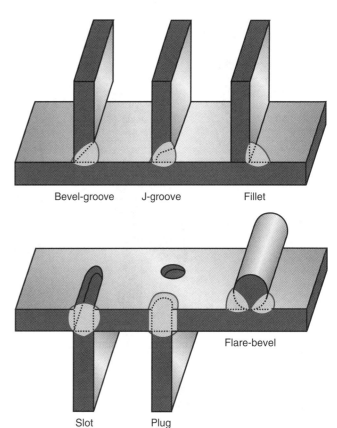

Goodheart-Willcox Publisher

Figure 6-5. Types of welds that can be made in a basic T-joint.

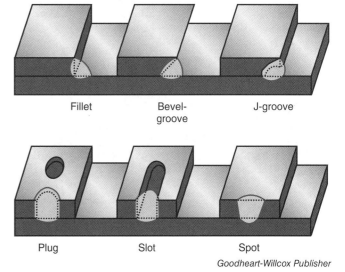

Goodheart-Willcox Publisher

Figure 6-6. Types of welds that can be made in a basic lap joint.

- V-groove butt weld.
- J-groove butt weld.
- U-groove butt weld.
- Flare-V-groove butt weld.
- Flare-bevel-groove butt weld.

T-joint. (Figure 6-5)
- Fillet weld.
- Plug weld.
- Slot weld.
- Bevel-groove weld.
- J-groove weld.
- Flare-bevel-groove weld.
- Melt-through weld.

Lap joint. (Figure 6-6)
- Fillet weld.
- Plug weld.
- Slot weld.
- Spot weld.
- Bevel-groove weld.
- J-groove weld.
- Flare-bevel-groove weld.

Corner joint. (Figure 6-7)
- Fillet weld.
- Spot weld.

- Square-groove weld or butt weld.
- V-groove weld.
- Bevel-groove weld.
- U-groove weld.
- J-groove weld.

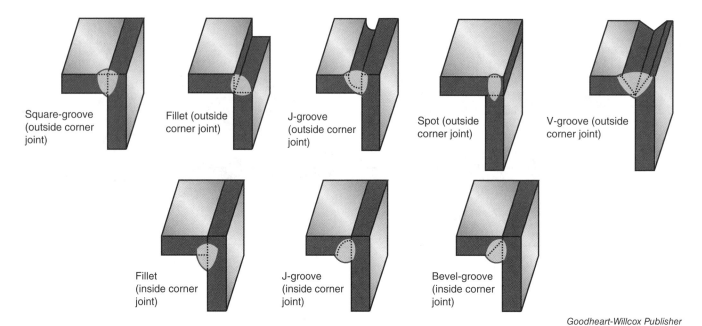

Goodheart-Willcox Publisher

Figure 6-7. Types of welds that can be made in a basic corner joint.

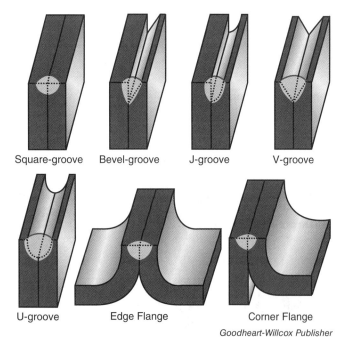

Goodheart-Willcox Publisher

Figure 6-8. Types of welds that can be made in a basic edge joint.

- Flare-V-groove weld.
- Edge weld.
- Corner-flange weld.
Edge joint. (Figure 6-8)
- Square-groove weld or butt weld.
- Bevel-groove weld.
- V-groove weld.
- J-groove weld.
- U-groove weld.

- Edge-flange weld.
- Corner-flange weld.

Double Welds

In some cases, a weld cannot be made from only one side of the joint. When a weld must be made from both sides, it is known as a *double weld*. See **Figure 6-9**. A V-groove weld made from both the top and bottom of the weld joint is considered a double V-groove weld. In the same manner, a fillet weld made on two sides of a T-joint is considered a double fillet weld.

Weldment Configurations

The basic joint often is changed to assist in a component's assembly. A weld joint might be modified to gain access to the weld joint or to change a weld's metallurgical properties. Some common weldment configuration designs are described in the following paragraphs.

Joggle-type joints are used in tanks (cylinder-to-head assemblies) where backing tooling or backing bars are not effective or cannot be used. See **Figure 6-10**. The automotive industry uses this joint type in the manufacture of unibody automobiles where one side of the panel must be flush. One side of the panel is joggled with special joggle tools. The unit is assembled, clamped, and the welding is completed as required.

The *tubular butt joint with built-in backing bar* design is used in tubular assemblies where tooling cannot be inserted into the pipe diameter or where overall dimensions are precise. In this case, the pipes are

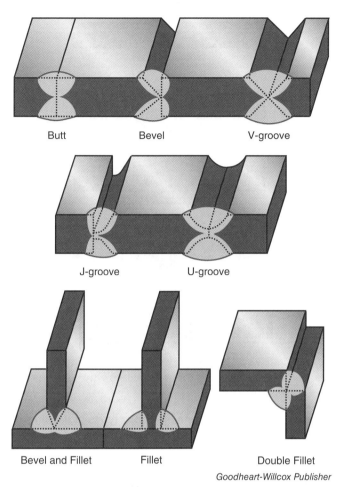

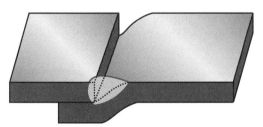

Figure 6-9. Common applications of double welds in basic joint designs.

Figure 6-10. Joggle-type joint.

Goodheart-Willcox Publisher

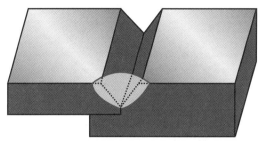

Goodheart-Willcox Publisher

Figure 6-11. Tubular butt joint with a built-in backing bar.

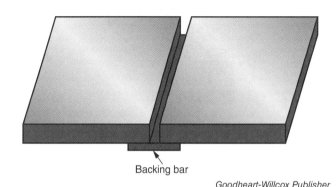

Backing bar

Goodheart-Willcox Publisher

Figure 6-12. Plate butt weld with a fabricated backing bar.

assembled until the lands (root faces) meet, controlling the overall length dimension. See **Figure 6-11**.

Another weldment design is the *butt joint with prefabricated backing bar*. A **backing bar** may be 1″ flat stock or larger, depending on the thickness of the base material. As the bar is assembled into the cylinder, tack welds are placed on one edge. After assembly of the mating part, the other side of the bar is tack welded. The purpose of the backing bar is to control the penetration of the weld bead on the backside of the joint. Since the weld usually penetrates into the backing bar, the bar is not removed after welding, **Figure 6-12**.

The major problem with this design is the fitup of the bar to the back of the base material. Any area that does not contact the back of the material will not control the heat flow, and penetration into the backing bar may not occur. Specially designed joints are used for controlling penetration into the joint where excessive penetration could cause a problem with liquid flow. See **Figure 6-13**.

Joining dissimilar metals often requires *buttering*—depositing a material on one of the joint pieces to make the joint materials metallurgically compatible with a common filler material. See **Figure 6-14**.

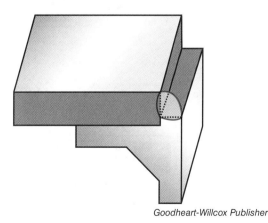

Goodheart-Willcox Publisher

Figure 6-13. Controlled weld penetration joint.

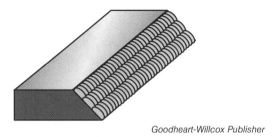

Goodheart-Willcox Publisher

Figure 6-14. Buttered weld joint face.

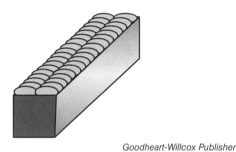

Goodheart-Willcox Publisher

Figure 6-15. Overlay welds protect the base material from wear or contamination.

When a material is required to protect the base metal from chemicals, heat, abrasion, or other forms of wear, a layer of weld metal is placed on the surface. Overlaying a weld in this manner is called *surfacing* or *cladding*, **Figure 6-15.**

Pipe joints often require a special backing ring for assembling and welding the joint. The type of material flowing through the pipe determines the ring design to be used. Because the weld metal penetrates the backing ring, it is not removable once welding is completed. See **Figure 6-16.**

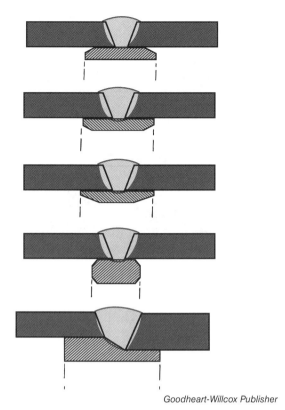

Goodheart-Willcox Publisher

Figure 6-16. Cross sections of various types of backing rings for pipe joints.

Welding Terms and Symbols

Communication from the weld designer to the welder is essential for proper completion of most weldments. See **Figure 6-17.** The American Welding Society developed the standard *AWS welding symbol*

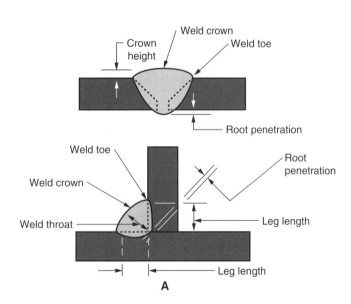

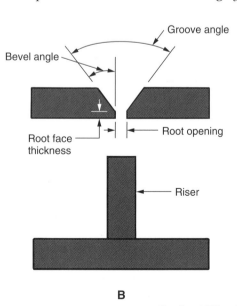

Goodheart-Willcox Publisher

Figure 6-17. Welding terminology. A—Terms that describe parts of the weld. B—Terms that describe the weld joint.

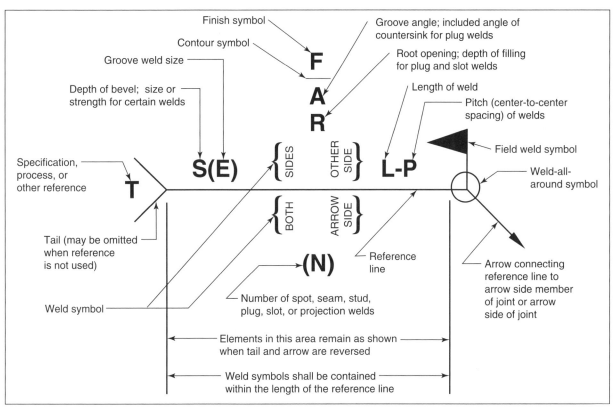

Figure 6-18. The AWS welding symbol conveys specific and complete information to the welder.

shown in **Figure 6-18**. This symbol is found on drawings to indicate the type of joint, placement, and the type of weld to be made. The symbol may also include other information, such as finish and contour of the completed weld.

It is important to study and understand each part of the welding symbol. **Figure 6-19** is a table showing basic *weld symbols* that are used with the AWS welding symbol to direct the welder in producing the proper weld joint. These symbols are standardized

Groove							
Square	Scarf	V	Bevel	U	J	Flare-V	Flare-bevel

Fillet	Plug or slot	Stud	Spot or projection	Seam	Back or backing	Surfacing	Edge

Figure 6-19. Basic weld symbols.

representations used for types and positions of welds, weld finishes, and welding techniques.

The arrow of the welding symbol indicates the point at which the weld is to be made. The line connecting the arrow to the *reference line* is always at an angle. The reference line is the horizontal line of the AWS welding symbol. All information about the weld to be made is positioned above, below, or on this line. Whenever the basic weld symbol is placed below the reference line, as shown in **Figure 6-20**, the weld is made on the side where the arrow points (referred to as the *arrow side*). Whenever the basic symbol is placed above the reference line, the weld is to be made on the *other side* of the joint, as shown in **Figure 6-21**. Dimensions placed on the symbol and drawings indicate the exact size of the weld. Study the examples of typical weld symbols and weldments shown in **Figure 6-22**.

The complete weld symbol gives the welder instructions for preparation of the base metal, the welding process to use, and the finish for the completed weld. Through careful use of these symbols, the weld designer can convey all the information needed to complete a weldment.

Classes that provide advanced study in the area of print reading for welders are recommended. Taking these classes helps welders improve their ability to read and interpret welding drawings. Studying print reading texts is also recommended. Only with a

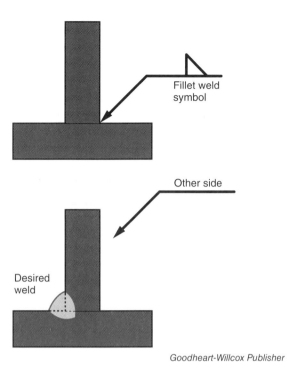

Goodheart-Willcox Publisher

Figure 6-21. The fillet weld symbol shown above the reference line indicates that the weld is located on the side of the joint opposite where the arrow points.

full understanding of the welding symbol and print reading should a welder attempt to fabricate any detailed weldment.

Weld Positions

Many weldments are made on site, so the welder must be able to weld in all positions—flat, horizontal, vertical, and overhead. The AWS designations for these weld positions (in order) are 1F, 2F, 3F, and 4F for fillet welds. In a similar manner, the AWS designations for groove welds are 1G, 2G, 3G, and 4G, and 1G, 2G, 5G and 6G for groove welds on pipe. See **Figure 6-23.** The effect of gravity on the molten weld pool differs with each position. Heat distribution also varies. These two factors make the skills for each position distinct. Practice is necessary to produce good welds in all positions. Keep these considerations in mind when welding in the various positions:

- The *flat* position has the greatest deposition rate. Flat welds usually have less porosity because the gas can rise to the top of the weld pool and escape before the metal solidifies.
- Undercut at the upper portion of the weld pool can be a major problem in the *horizontal* position. *Undercut* is a groove melted into the base metal

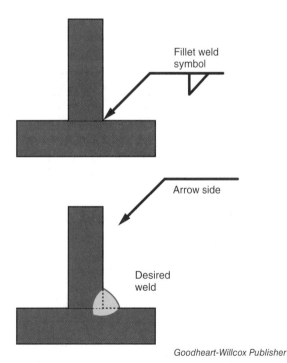

Goodheart-Willcox Publisher

Figure 6-20. The fillet weld symbol shown below the reference line indicates that the weld is located on the side at which the arrow points.

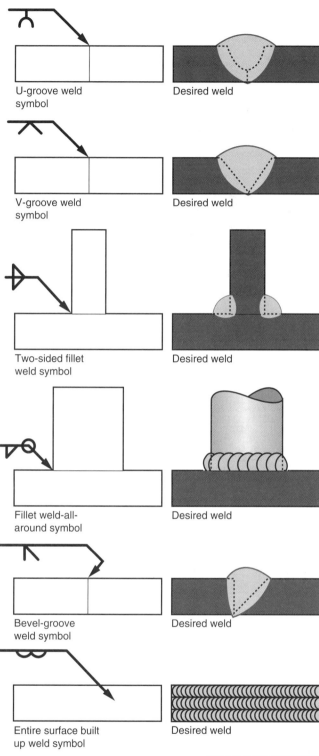

U-groove weld symbol

Desired weld

V-groove weld symbol

Desired weld

Two-sided fillet weld symbol

Desired weld

Fillet weld-all-around symbol

Desired weld

Bevel-groove weld symbol

Desired weld

Entire surface built up weld symbol

Desired weld

Goodheart-Willcox Publisher

Figure 6-22. Typical weld symbols and applications.

next to the toe of the weld and left unfilled by weld metal.

- Heat rises during *vertical* welding, requiring close observation of the molten pool to prevent sagging and overheating of the weld.

- *Overhead* welding is the most tiring of all positions. It also has the slowest metal deposition. Making weave beads is extremely difficult due to metal sagging. Overhead welds are prone to porosity.

Always wear protective clothing and leathers to shield you from falling sparks and metal. Do not stand directly under the spark stream. Wear protective eye lenses. Never clean the slag on a weld without eye protection. Stay aware of your environment when your hood is down and you are welding. If flammable material is present, take measures to prevent fires. Keep a fire extinguisher at your work area.

Design Considerations

Design of the weld type and weld joint is of prime importance if the weldment is to do the intended job. The weld should be made at reasonable cost. Weld design considerations include the following:
- Base material type and condition (annealed, hardened, tempered).
- Service conditions (pressure, chemical, vibration, shock, wear).
- Physical and mechanical properties of the completed weld and heat-affected zone. The heat-affected zone (HAZ) is the area around the weld joint that has been changed, mechanically or chemically, by the heat of the welding operation.
- Preparation and welding cost.
- Assembly configuration and weld access.
- Welding equipment and tooling.

Butt Joint and Welds

Butt joints are used when strength is required. These joints are reliable and can withstand stress better than any other type of weld. To achieve full stress value, the weld must have 100% penetration. This can be accomplished by completely welding through from one side. The alternative is to weld both sides with the weld joining in the center.

Thin gauge materials are more difficult to fit up for welding. They require costly tooling to maintain proper joint configuration. Tack welding can be used to hold the components during assembly. However, tack welds present several problems:
- Conflicting with the final weld penetration into the weld joint.
- Adding to the final crown dimension (height).
- Cracking during the welding operation due to heating and expansion of the joint.

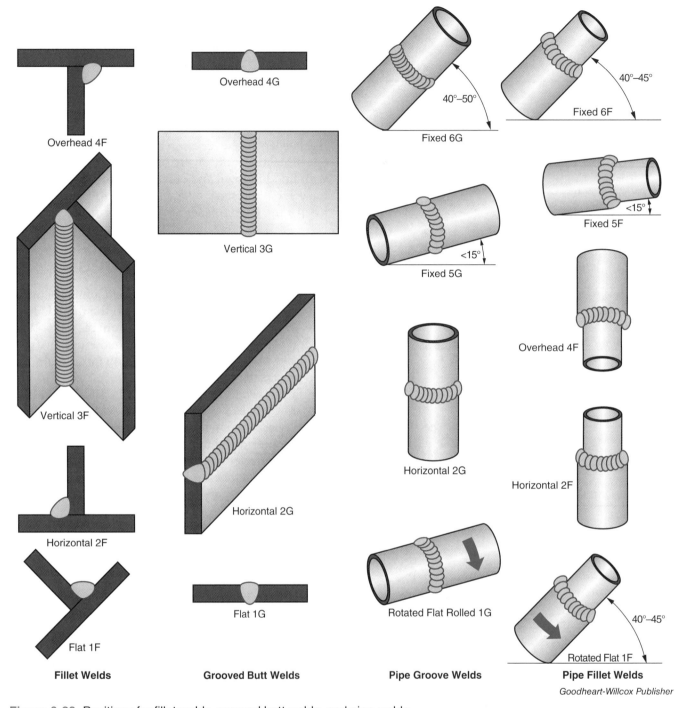

Figure 6-23. Positions for fillet welds, grooved butt welds, and pipe welds.

- Contaminating the weld joint when cleaning the tack weld slag, causing weld defects.

The expansion of the base material during welding can result in mismatch, **Figure 6-24**. When mismatch occurs, the weld does not completely penetrate the joint. Many specifications limit highly stressed joints to a maximum mismatch of 10% of the joint thickness.

Whenever possible, butt joints should mate at the bottom of the joint, **Figure 6-25**. Joints of unequal

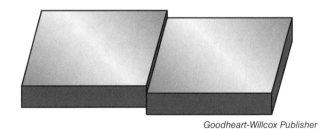

Goodheart-Willcox Publisher

Figure 6-24. When under stress, welds made with mismatched joints often fail below the rated load.

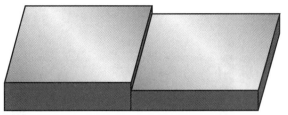

Figure 6-25. Mating the joint at the bottom equalizes the load under stress. The weld is placed on the top and penetrates completely through the joint.

thicknesses should be machined in the weld area to provide even surfaces for adequate fusion, **Figure 6-26**. Where this cannot be done, the heavier piece can be machined on the upper part of the joint.

Weld Shrinkage

Butt welds always shrink across (transverse) the joint during welding. A shrinkage allowance must be made if the postwelding dimensions have a small tolerance or if the outer edges of the material are to be trimmed to a final dimension. Butt welds on pipe, tubing, and cylinders also shrink, **Figure 6-27**. When dimensions must be held, a shrinkage test can be done, **Figure 6-28**. Additional material can be added to the

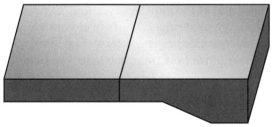

Figure 6-26. Joints of unequal thickness absorb different amounts of heat and expand at different ratios. Equalize heat flow by tapering the heavier material to the thickness of the thinner material.

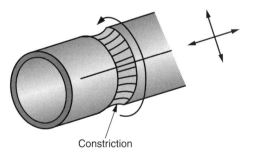

Constriction

Figure 6-27. During welding, butt welds shrink both transversely and longitudinally.

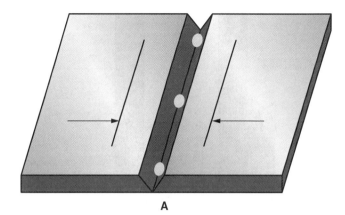

A

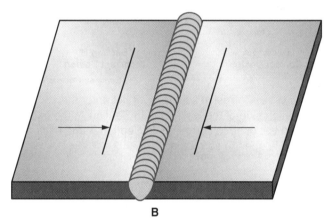

B

Figure 6-28. Weld joint shrinkage can be determined in four steps. A–1. Tack the test joint together. 2. Scribe parallel lines on approximately 2″ centers. Record this dimension. B–3. Weld the joint with the test weld procedure. 4. Measure the new distance between the scribed lines.

overall length of the part for final trimming. Heavier materials shrink more than thinner materials.

Lap Joints and Welds

Lap joints require little preparation and can be single-fillet, double-fillet, plug, slot, or spot-welded. These joints are used in static-load applications or for the repair of unibody automobiles. If the joint will be exposed to corrosive liquids, such as a weld seam in a chemical storage tank, both edges of the joint must be welded. See **Figure 6-29**. One of the major problems with lap joint design occurs when the component parts are not in close contact, as they should be for a fillet weld. See **Figure 6-30**. A bridging fillet weld that leads to incomplete fusion at the root of the weld and an over-size fillet weld dimension must be made. When a lap joint in sheet or plate material has a gap due to poor fitup, adequate clamps or tooling must be used to maintain contact of the material at the weld joint.

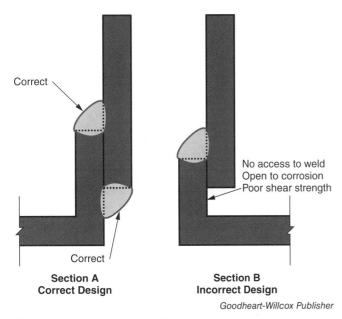

Figure 6-29. Corrosive liquids must not enter the penetration side of the weld joint. In Section A, the back of the weld is closed to corrosion. In Section B, the back of the weld is open to corrosion.

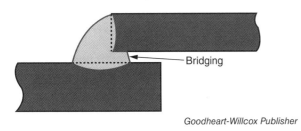

Figure 6-30. Lap joint problem areas are often the result of improper fitup.

In the assembly of cylindrical parts, an *interference fit* of the mating parts eliminates the problem of excessive gaps between parts, **Figure 6-31**. The inside diameter of the outer part is made several thousandths of an inch smaller than the outside diameter of the inner part. Before assembly, the outer part is heated until it expands enough to slide over the inner part. As it cools, it locks onto the smaller part. Small tack welds can be used to hold the assembly together during welding.

T-Joints and Welds

T-joints join parts at angles to each other. T-joints may be single-fillet, double-fillet, or groove-and-fillet welds, **Figure 6-32**. Fillet weld sizes must conform to allowable design loads. AWS specifies where fillet welds can be used, as well as the minimum and maximum sizes permissible in the construction of buildings and bridges. When welding on materials

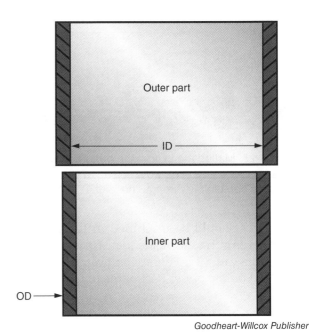

Figure 6-31. In an interference fit of mating cylindrical parts, the ID of the outer part is made several thousandths of an inch smaller than the OD of the inner part.

less than 1/4″ thick, and design loads are unknown, follow this rule of thumb—the fillet weld lengths must equal the thickness of the thinner material being joined. See **Figure 6-33**.

The main problem in making fillet welds is lack of penetration at the joint intersection. To prevent this condition, make *stringer beads* (welds made without side-to-side motion) at the intersection. Weave beads across the intersection are prone to lack of penetration.

Corner Joints and Welds

Corner joints are similar to T-joints. They consist of sheets or plates mating at an angle to one another, **Figure 6-34**. Corner joints are usually used in conjunction with groove welds and fillet welds. When thin materials are used, the assembly of component parts may be difficult without proper tooling. Tack welding and welding without extensive tooling often causes distortion and buckling of thinner materials. For the most part, these types of joints should be limited to heavier materials in structural assemblies.

Edge Joints and Welds

Edge welds are used when the edges of two sheets or plates are adjacent and have approximately parallel planes at the point of welding. See **Figure 6-35**. Edge welds are applicable on thinner gauges of metal and are not used in structural assemblies. Only full penetration welds should be used with stress or pressure applications.

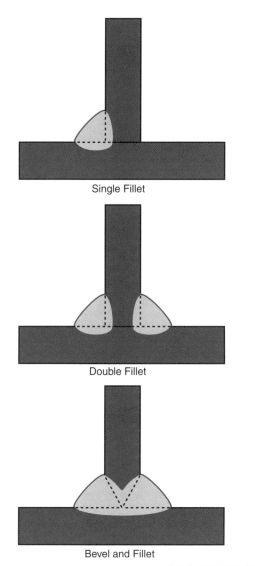

Single Fillet

Double Fillet

Bevel and Fillet

Goodheart-Willcox Publisher

Figure 6-32. Types of T-joints and welds.

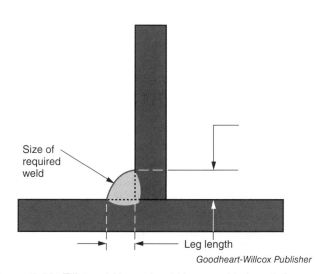

Size of required weld

Leg length

Goodheart-Willcox Publisher

Figure 6-33. Fillet weld legs should be equal in length from the root of the joint. Unequal leg length, unless specified, will not carry the designed load and may fail when stressed.

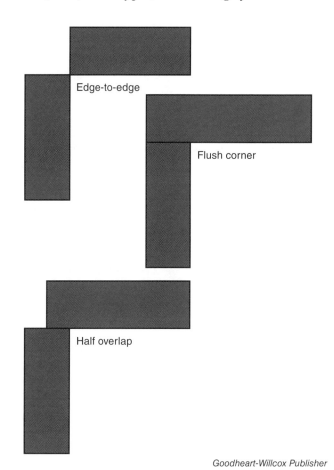

Edge-to-edge

Flush corner

Half overlap

Goodheart-Willcox Publisher

Figure 6-34. Common corner-joint designs used in the fabrication of component parts.

Special Designs and Procedures

Special designs and procedures are required in the fabrication of a weldment under these circumstances:

- Welds cannot be thermally treated due to their configuration or size. Additional material is added to the joint thickness, and the weld is made, restricting the heat flow into the thinner material. See **Figure 6-36**. This type of joint achieves the full mechanical values of the base material.
- Joining dissimilar metals often requires buttering one material to allow the joint to be welded with a common filler material. See **Figure 6-37**.
- Chill (backing) bars and tooling can be used to localize and remove heat during the welding application. See **Figure 6-38**.
  Other benefits of tooling include the following:
- Aligning components before and during the welding operation.
- Checking component parts for proper fitup.
- Preventing excessive drop-through penetration on the other side of the weld.

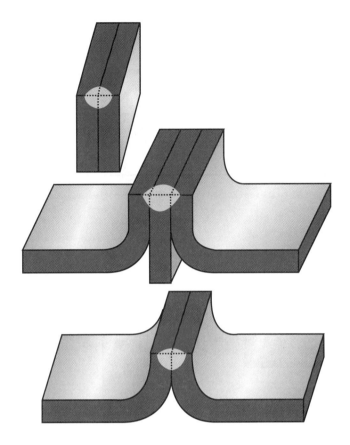

Figure 6-35. Common edge-joint designs.

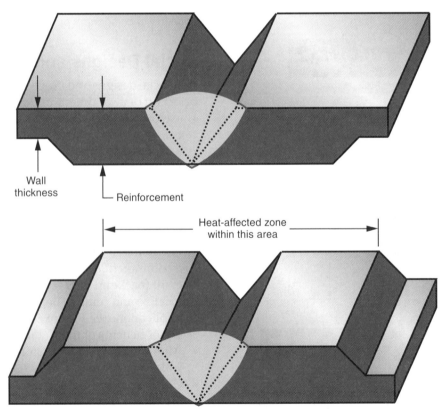

Wall
thickness

Reinforcement

Heat-affected zone
within this area

Figure 6-36. Joint thickness and filler material tensile strength combined are equivalent to the strength of the base material.

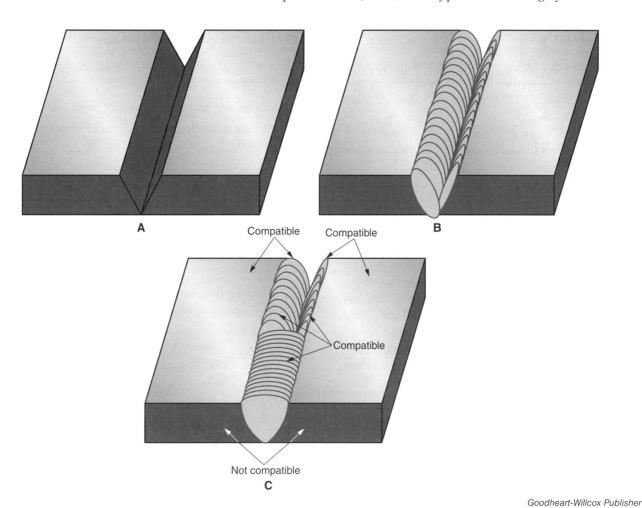

Figure 6-37. Buttering is commonly used to adapt dissimilar metals for welding.

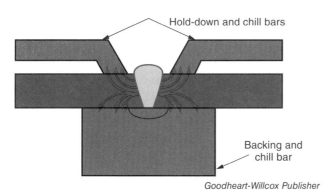

Figure 6-38. Tooling and chill bars remove heat from the weld and resist heat flow into the base material.

Summary

- All welds used to join two pieces of material can be classified as either fillet welds or groove welds.
- The five basic weld joints are butt, T, edge, corner, and lap joints.
- The American Welding Society developed the welding symbol as a standard means of showing the proper weld configuration on a print. This symbol indicates the type of joint, placement, and the type of weld to be made.
- The four positions of the weld joint are flat, horizontal, vertical, and overhead.
- Design of the weld type and weld joint is of prime importance if the weldment is to do the intended job. Factors to be considered are the material type and condition, service conditions of the weld, physical and mechanical properties of the completed weld and heat-affected zone, preparation and welding cost, assembly configuration and weld access, and the welding equipment and tooling available.

Review Questions

Answer the following questions using the information provided in this chapter.

1. Name the five basic types of weld joints.
2. Welds made from both sides of the joint are called _____ welds.
3. *True or False?* Joggle-type joints are used with backing bars.
4. Joining dissimilar metals often requires *buttering*. What does this term mean?
5. On welding drawings, the symbol above the reference line indicates the weld is to be made on the _____ of the joint.
6. List the four basic welding positions defined by the American Welding Society.
7. The effect of gravity on the molten weld pool and _____ distribution differs between the four basic welding positions.
8. Undercut at the upper portion of the weld pool can be a major problem in the _____ welding position.
9. When mismatch occurs, the weld does not completely _____ the joint.
10. *True or False?* Butt joints should mate at the bottom of the joint whenever possible.
11. *True or False?* Butt welds always shrink across the joint during welding.
12. What problem does an interference fit of mating parts eliminate?
13. To eliminate the problem of lack of penetration at the joint intersection of a fillet weld, make _____ beads at the intersection.
14. Edge welds are applicable on _____ gauges of metal and are not used in structural assemblies.
15. Bars that are used during welding to localize and remove heat from the weld area are called _____ bars.

Gas Metal Arc Welding Procedures and Techniques

Objectives

After studying this chapter, you will be able to:

❏ Adjust welding current and voltage parameters to produce desired weld qualities.

❏ Vary electrode extension, travel speed and direction, and gun angle to produce desired weld qualities.

❏ Distinguish between bead patterns.

❏ Determine values for a weld schedule based on results from test welds.

Technical Terms

backhand welding
burn-through
downhill welding
electrode extension
forehand welding
pull welding
push welding
stickout
stringer bead pattern
travel angle
uphill welding
weave bead pattern
welding voltage
weld schedule
wire feed speed
work angle

Welding Variables and Parameters

The GMAW process can be either a semiautomatic or a fully automatic operation. In fully automatic operation, the machine controls all of the parameters and variables. In the semiautomatic operation, the hands-on skill of the welder significantly affects the final quality of the weld. The welder must be able to set up the machine properly and skillfully operate the gun in order to create a weld that meets the quality requirements of the fabrication specification. Welding variables that are determined after the material type and thickness and mode of welding have been selected are described in the following sections.

Welding Current/Amperage (Wire Feed Speed)

With a constant voltage machine, the welding current is established by the *wire feed speed*. The wire feed speed is the rate at which the welding wire is fed through the welding gun. Increasing the wire feed speed increases welding current, while decreasing wire feed speed decreases welding current.

Wire feed speed is measured in *inches per minute (ipm)*. Determine wire feed speed as follows:

1. Turn the machine on and run the wire to the end of the gas nozzle or the contact tip.
2. Depress the arc start switch (this starts wire feed) and run for 10 seconds.
3. Measure wire length from the end of the contact tip or the gas nozzle, **Figure 7-1**.

Goodheart-Willcox Publisher

Figure 7-1. To establish wire speed, carefully measure the length of wire fed in 10 seconds. Multiply the length by 6 for wire speed inches per minute. Wire speed directly affects the welding current.

4. Multiply wire length by six (10 seconds = 1/6 of a minute). The resulting number is the wire speed in inches per minute.

Keep the end of the wire away from the work or disconnect the work clamp and make sure it is safely out of the way. If contact is made, an arc will occur.

Welding Voltage

The *welding voltage* determines the arc gap established between the end of the electrode and the workpiece. This parameter is set on the welding machine. The machine automatically changes the amperage output to maintain this preset arc length. Charts listing the initial welding parameters for steel, stainless steel, and aluminum are located in the *Reference Section*.

Electrode Extension

Electrode extension is the distance from the contact tip to the end of the electrode. It is important to measure the actual distance from the contact tip to the end of the electrode and not just the amount of electrode extending from the end of the gas nozzle. The term *stickout* refers to the length of unmelted electrode extending past the end of the gas nozzle. The dimensions used for these terms are shown in **Figure 7-2**.

The welder controls electrode extension through the handling of the gun while welding. Changing the extension from the originally established dimension has a marked effect on the melting of the elec-

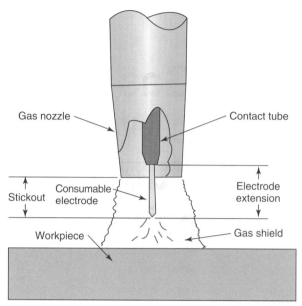

Goodheart-Willcox Publisher

Figure 7-2. The actual electrode extension is from the end of the contact tip to the end of the electrode. The term *stickout* describes the distance from the nozzle to the end of the electrode wire.

trode. Increasing the extension causes more resistance heating of the wire and a lower welding amperage, which is useful when welding on gapped joints to prevent **burn-through**. Burn-through is caused by excessive amperage for the thickness of the material to be welded. Conversely, decreasing the extension makes less wire available for resistance heating. This increases amperage for greater penetration, **Figure 7-3**

Travel Speed

The travel speed of the welding gun must be regulated to produce a good weld. In machine welding, the speed is set in inches per minute (ipm). In manual

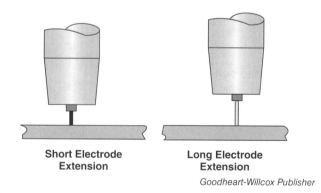

Goodheart-Willcox Publisher

Figure 7-3. Electrode extension variations. Use these variations from the welding procedure dimensions to increase or decrease welding heat into the molten weld pool.

welding, the welder controls this variable. The welder must adapt the speed to changing shapes, improper fitup, gaps, and other variables as the weld continues. The welder's skill is important to the quality of the weld.

Direction of Travel

The GMAW operation can be done with the welding gun pointing back at the weld and the weld progressing in the opposite direction. This is called *backhand welding* or *pull welding*. See **Figure 7-4**. Advantages of backhand welding include a more stable arc, less spatter, and deeper penetration.

The other method of gun manipulation is to point the gun forward in the direction of travel. This method is called *forehand welding* or *push welding*. This is the recommended direction for aluminum and when the spray transfer or pulsed-spray transfer mode of deposition is used. See **Figure 7-5**. In compar-

ison to backhand welding, forehand welding results in more spatter, less penetration, and better visibility of the weld seam. This technique provides increased cleaning action of the base metal, which is why it is used for welding aluminum.

Gun Angle

Gun angles are defined in either longitudinal (along the weld) or transverse (across the weld) angle dimensions. The longitudinal angle is referred to as the *travel angle*. Longitudinal angles generally range between 10° and 25° off perpendicular. The transverse angle is referred to as the *work angle*. See **Figure 7-6**. Transverse angles range from 0° to 45°, depending on the welding position and joint type.

Gun angle affects weld penetration, bead form, and final weld bead appearance. As the weld progresses, changes in gun angle may be required to maintain good weld pool control and weld bead shape.

Figure 7-7 shows the transverse gun angles (for initial welding setup only) and bead placement for flat multi-pass groove welds. Longitudinal angles are 10°–15° from the vertical position. Each variation in groove angle or thickness dimension affects the required gun angle. Remember, the molten weld pool must flow into the previous pass and the side walls of

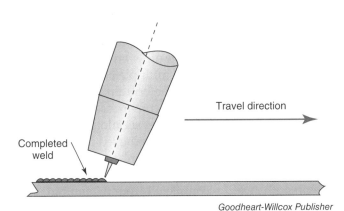

Goodheart-Willcox Publisher

Figure 7-4. The backhand (pull welding) technique requires more skill on the part of the welder than the forehand technique, since the weld joint is difficult to see because of the position of the gas nozzle.

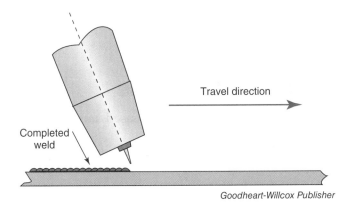

Goodheart-Willcox Publisher

Figure 7-5. The forehand or push welding technique is easy to use because the weld joint is directly in front of the electrode. Compared to backhand welding, there will be less penetration into the joint and spatter will increase.

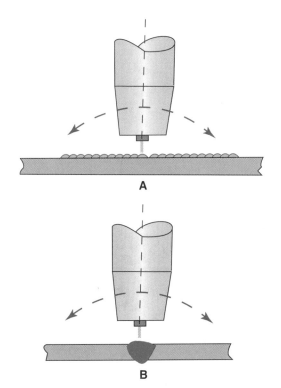

Goodheart-Willcox Publisher

Figure 7-6. Gun angles. A—Travel angle. B—Work angle.

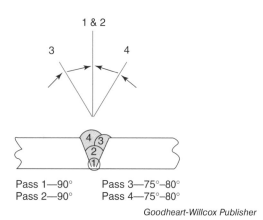

Pass 1—90°   Pass 3—75°–80°
Pass 2—90°   Pass 4—75°–80°

Goodheart-Willcox Publisher

Figure 7-7. Weld gun angles and bead placement for multipass groove welds in the flat position. Longitudinal angles are 10°–15° from the vertical position. The angles shown are initial transverse angles and may need to be adjusted for proper bead placement, depending on the joint's bevel angles.

the joint. Develop your welding skill so you can properly fill the groove joint with a sound weld.

Figure 7-8 shows the transverse gun angles (for initial setup only) and bead placement for horizontal multi-pass bevel groove and V-groove welds. Longitudinal angles are 10°–15° from the perpendicular position. Remember to make the weld beads smaller when welding out of position and always make stringer beads, not weave beads. Fill each layer from the bottom upward.

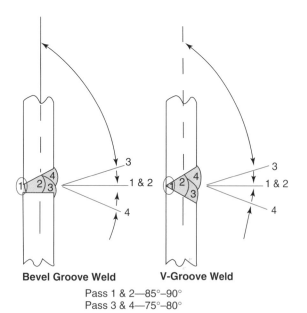

Bevel Groove Weld   **V-Groove Weld**

Pass 1 & 2—85°–90°
Pass 3 & 4—75°–80°

Goodheart-Willcox Publisher

Figure 7-8. Horizontal groove-weld gun angles and bead placement. The initial gun transverse angles shown may need to be adjusted for proper bead placement, depending on the joint's bevel angles.

Pass 1—45°
Pass 2—35°
Pass 3—55°

Goodheart-Willcox Publisher

Figure 7-9. Horizontal fillet weld gun angles and bead placement. Longitudinal angles are 10°–15° from the vertical position. The angles shown are initial transverse angles and may need to be adjusted based on the number of passes and the bead contour desired.

Figure 7-9 shows the transverse gun angles (for initial welding setup) and bead placement for horizontal multipass fillet welds. For a single pass weld, use the 45° angle and use the other angles for a multiple pass weld. On a horizontal multiple pass weld, always make the second pass on the bottom. The completed weld should have a slightly convex crown and equal length legs. The size of the fillet should be equal to the thickness of the thinnest material used in the joint.

Figure 7-10 shows the gun angles for *uphill welding* (commonly called *vertical-up*). In uphill welding, the weld is completed from the bottom of the joint upward to the top of groove and fillet welds. For fillet welds, the bead placement is the same as for flat and horizontal welds. Weave beads can be used after the root pass is made and if the joint access is sufficient to accept the additional metal. **Figure 7-11** shows the gun angles for *downhill welding* (commonly called *vertical-down*). In downhill welding, the weld is completed from the top of the weld joint to the bottom of groove and fillet welds.

Weld Bead Patterns

The two types of patterns used for depositing metal are as follows:

- **Stringer bead pattern.** In the *stringer bead pattern,* travel is along the joint with very little side-to-side motion. It may be made with a small zigzag motion or in a small circular motion. See **Figure 7-12.** As you practice these motions, vary your forward speed as necessary to maintain an even bead contour and shape.

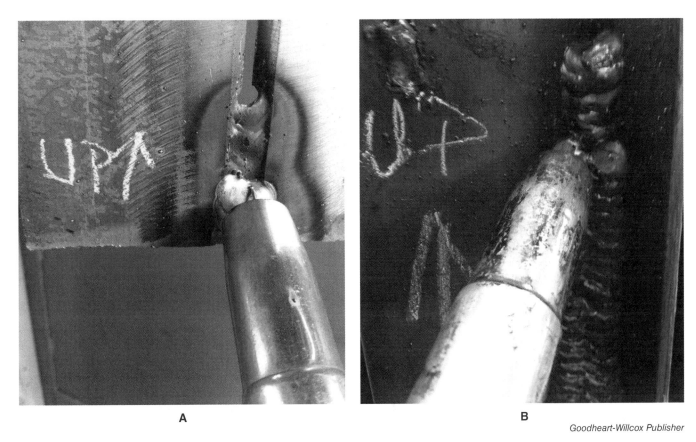

A   B

Goodheart-Willcox Publisher

Figure 7-10. Gun placement for welding uphill. A—Groove weld. B—Fillet weld.

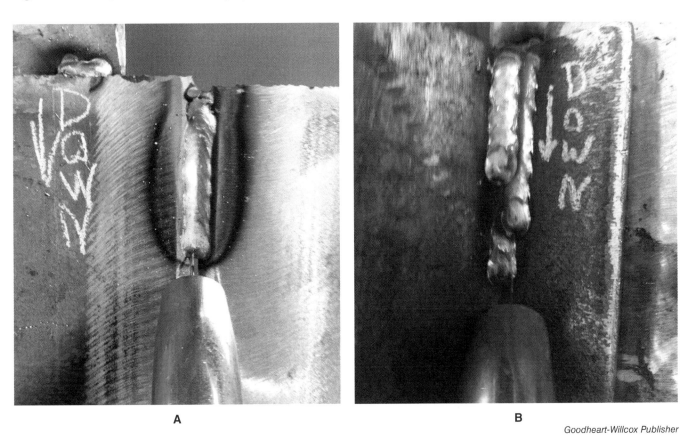

A   B

Goodheart-Willcox Publisher

Figure 7-11. Vertical groove and fillet weld gun placement when welding downhill (vertical down). A—Groove weld. B—Fillet weld.

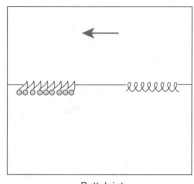

Butt Joint

Goodheart-Willcox Publisher

Figure 7-12. Stringer bead patterns. The bead on the left is called a *backstep* or *zigzag* technique and requires a dwell point on each swing backward. Dwell points are marked with dots. The circular bead technique on the right does not have any dwell points.

- **Weave bead pattern.** The *weave bead pattern* is also called a *wash bead pattern* or an *oscillation bead pattern*. The term *oscillation* refers to side-to-side movement. The weld is wider than a stringer bead and requires a dwell (wait) of the gun at the end of each weave to fill the metal into the weld without undercut. See **Figure 7-13**.

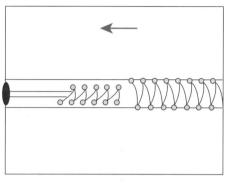

V-Groove Joint

Goodheart-Willcox Publisher

Figure 7-13. Weave bead patterns. The weave bead pattern is prone to undercut at the outer edges of the weld and requires a hold or dwell point as indicated by the dots.

Weld Schedules

The welding of component parts requires a combination of many parameters and variables. See **Figure 7-14**. To document all the materials, wire sizes, types of gases, and techniques used to make a weld, these items are listed on a form called a *weld schedule*, **Figure 7-15**.

GMAW Welding Variables								
Welding Variables to Change	Penetration		Deposition Rate		Bead Size		Bead Width	
	Increase	Decrease	Increase	Decrease	Increase	Decrease	Increase	Decrease
Wire Feed Speed— Amperage	↑	↓	↑	↓	↑	↓	No effect	No effect
Voltage	Little effect	Little effect	No effect	No effect	No effect	No effect	↑	↓
Travel Speed	Little effect	Little effect	No effect	No effect	↓	↑	↑	↓
Wire Stickout	↓	↑	↑	↓	↑	↓	↓	↑
Wire Size	↓	↑	↓	↑	No effect	No effect	No effect	No effect
Gas % of CO₂	↑	↓	No effect	No effect	No effect	No effect	↑	↓
Torch Angle	Backhand	Forehand	No effect	No effect	No effect	No effect	Backhand	Forehand

↑ Increase

↓ Decrease

Goodheart-Willcox Publisher

Figure 7-14. Adjusting welding parameters and techniques affects the weld.

JM WELDING LLC
PRODUCTION WELDING SCHEDULE

Process	_____	Job #	_____
WPS	_____	PQR	_____
Weld Type	_____	Position	_____
Base Metal	_____	Thickness	_____
Electrode	_____	Diameter	_____
Wire Speed	_____ IPM	Amperage Range	_____
Electrode Stickout	_____	Voltage Range	_____
Inductance	_____	Nozzle Diameter	_____
Shielding Gas	_____	Flow Rate	_____ CFH

Travel Direction

Forehand (Push)	_____	Backhand (Pull)	_____
Uphill	_____	Downhill	_____
Preweld Cleaning	_____	Postweld Cleaning	_____

Preheat Temp. _____   Interpass Temp. _____   Postheat Temp. _____

Welder #	_____	Quality Control #	_____
Quantity	_____	Quantity Checked	_____
Date	_____	Date	_____

Notes   Special Instructions   Part Sketch

Figure 7-15. Weld schedules are used by industry to record data for ready reference for each welding job.

The first step is to select an initial setting from the appropriate chart in the *Reference Section* for the type and thickness of material involved. Select voltage and wire feed speed based on the electrode wire size. Record the machine settings and techniques that will be used as a starting point for the weld test on the weld schedule.

The next step is to clean the material and make test welds. During the test weld period, listen to the sound of the arc as you are welding. When welding is performed in the short circuiting transfer mode, the molten wire should leave the end of the wire evenly without excessive noise and popping as it contacts the weld pool. A buzzing sound indicates the arc voltage and wire feed are properly set. When welding is performed in the spray transfer mode, there should be a continuous hissing sound without popping and interruption. During the test, make adjustments as needed. If the test produces an acceptable weld, revise the weld schedule to reflect the adjusted values.

Summary

- After the type and thickness of material and transfer mode are determined, variables to be established include wire feed speed (amperage), voltage, electrode extension, travel speed, and travel direction.
- The welding voltage determines the arc gap established between the end of the electrode and the workpiece.
- Increasing wire feed speed increases welding current, and decreasing feed speed decreases welding current.
- Increasing electrode extension causes more resistance heating of the wire and a lower welding amperage. Decreasing the electrode extension has the opposite effect.
- Advantages of backhand welding are a more stable arc, less spatter, and deeper penetration. Forehand welding results in more spatter, less penetration, and better visibility of the weld seam.
- Gun angle affects weld penetration, bead form, and final weld bead appearance.
- The two types of patterns used for depositing metal are the stringer bead pattern and weave bead pattern.
- A weld schedule is a document listing the variables required to produce a desired weld.

Review Questions

Answer the following questions using the information provided in this chapter.

1. With a constant voltage machine, the _____ establishes the welding current.
2. The arc gap between the end of the electrode and the workpiece is determined by the welding _____.
3. Increasing the electrode extension causes more resistance heating of the wire and a(n) _____ welding amperage.
4. What does the term *stickout* refer to?
5. List three advantages of backhand welding.
6. _____ angles generally range between 10° and 25° off perpendicular.
7. The weave bead pattern requires a(n) _____ at the end of each weave to fill the metal into the weld without undercut.
8. Downhill welding is commonly referred as _____.
9. A weld bead pattern made with side-to-side motion is commonly called a(n) _____ bead.
10. A weld _____ contains information about the weld joint design and the welding parameters and variables.

Gas Metal Arc Welding Procedures for Carbon Steels

8

Objectives

After studying this chapter, you will be able to:

❑ Explain the difference between various types of carbon steels.

❑ Select the appropriate electrode wire for specific welding conditions.

❑ Prepare a carbon steel joint for welding.

❑ Apply preheat, interpass temperature, and postheat in correct proportions on various joints.

❑ Describe tooling requirements for the GMAW process.

❑ Establish a GMAW procedure for welding carbon steel.

❑ Understand the procedures for plug and slot welding carbon steels.

Technical Terms

arc blow
carbon steels
chrome-moly steels
deoxidizer
free machining steels
high-carbon steel
interpass temperature
low-alloy steels
low-carbon steel
medium-carbon steel

metal-cored electrode wire
postheating
preheating
stress relief

Carbon Steels

Many types of metals are used in welding operations. The steel family includes various types and grades of carbon steels. *Carbon steels* are a group of steels that contain carbon, manganese, and silicon, along with small amounts of other elements. These steels are magnetic and reach their melting point at approximately 2500°F (1371°C). This group of steels contains the following elements:

- Carbon—1.70% maximum.
- Manganese—1.65% maximum.
- Silicon—0.60% maximum.

Carbon steels can be further classified depending on the percentage of carbon they contain:

- *Low-carbon steel*—up to 0.14% carbon.
- *Medium-carbon steel*—0.15% to 0.29% carbon.
- *High-carbon steel*—0.30% to 0.59% carbon.

The carbon content of steel affects its physical qualities. As the percentage of carbon in steel increases, the steel becomes harder but also more brittle and difficult to weld. Some steels have been modified for use as machining stock by adding special materials to permit the machining operation. These steels are called *free machining steels* or may have other names denoting their use. These materials are not weldable.

Low-Alloy Steels

Low-alloy steels contain varying amounts of carbon in addition to many alloying elements. These elements include chromium, molybdenum, nickel, vanadium, and manganese. The alloying elements increase the strength, toughness, and in some cases, the corrosion resistance of the steel. Prior to welding these materials either alone or in combination, the welding procedure should be qualified in conjunction with the thermal procedures used during welding and the heat treatment to be completed after welding. Qualification of welding procedures is discussed in Chapter 17.

Quenched and Tempered Steels

Quenched and tempered steels and high-strength steels (HSS) are produced by a number of steel companies. They have many trade names. These steels are hardened and tempered at the mill for specific mechanical values. With these processes, the strength, impact resistance, corrosion resistance, and other properties can be improved. Before welding this type of steel, the manufacturer should be consulted for the proper procedure. This ensures that the metal's special attributes are maintained.

Chrome-Moly Steels

The alloys known as the *chromium-molybdenum steels* (*chrome-moly steels*) are used in applications requiring high strength. These steels can be annealed, hardened, or tempered. Special procedures are required when welding these materials to yield the desired mechanical properties. Always qualify the weld joint design, filler material type, and the welding procedure by adequate testing prior to welding this material.

Electrodes for Carbon Steel

Electrode wires used for welding mild steels are defined in the American Welding Society specification AWS A5.18 *Specification for Carbon Steel Electrodes and Rods for Gas Shielded Arc Welding.* Codes identify different electrode materials. An example of an electrode wire number is *ER70S-2*. The letter *E* stands for electrode. The *R* stands for rod, meaning the electrode wire can also be used for gas tungsten arc welding (GTAW) applications. The number *70* means the wire can produce a weld with 70,000 pounds of tensile strength. The letter *S* identifies the wire as a solid electrode wire. The number *2* identifies the class of wire.

Each class of wire is defined by the properties the wire will produce when welding is performed with a specific single shielding gas or combination of gases. This number also identifies the type of deoxidizer used in the electrode. A major consideration when selecting the electrode is the material cleanliness—the more oxidation and mill scale, the higher the amount of deoxidizer necessary. A *deoxidizer* is

an element that helps remove oxygen and nitrogen from the weld, reducing the occurrence of weld metal porosity. For carbon steel, the deoxidizers typically used are manganese and silicon. Other elements used to a lesser degree are aluminum, zirconium, and titanium. The higher the silicon deoxidizer level, the more fluid the weld pool becomes, thus helping the wetting action at the weld toe and influencing the bead appearance.

The chemical compositions of various carbon steel solid wires are shown in the *Reference Section*. Solid carbon steel electrodes with deoxidizing alloys include the following:

- **ER70S-2.** This wire is heavily deoxidized, meaning that it contains agents to fight oxidation. It is designed to produce sound welds in all grades of carbon steel—killed, semikilled, and rimmed. Due to added deoxidants (aluminum, zirconium, and titanium), this wire can weld carbon steel with a rusty surface. Argon-oxygen, argon-carbon dioxide, and carbon dioxide shielding gases can be used. In general, an extremely viscous (not fluid) weld pool is produced. This makes ER70S-2 an ideal wire for short circuiting transfer welding out-of-position. A high oxygen or carbon dioxide content improves the wetting action of the pool.

- **ER70S-3.** This wire is widely used with GMAW. ER70S-3 wires can be used with carbon dioxide, argon-oxygen, or argon-carbon dioxide shielding gases for welding killed and semikilled steels. This electrode contains medium levels of silicon and manganese. Rimmed steels should be welded with only argon-oxygen or argon-carbon dioxide shielding gases. High welding currents used with carbon dioxide shielding gas may result in low strength. Either single-pass or multipass welds can be made with this electrode wire. The tensile strength for a single-pass weld in thin-gauge, low- and medium-carbon steels exceeds the base material while ductility is adequate. In a multipass weld, the tensile strength ranges between 65,000 and 85,000 psi depending on the base metal dilution and type of shielding gas. The weld pool is more fluid than that of the ER70S-2. The ER70S-3 has better wetting action and flatter beads. This wire is commonly used on automobiles, farm equipment, and home appliances.

- **ER70S-4.** This wire contains a higher level of silicon and manganese than the ER70S-3. This composition improves the soundness of semikilled steels and increases weld metal strength. ER70S-4 performs well with argon-oxygen, argon-carbon dioxide, and carbon dioxide shielding gases. It also can be used with either spray transfer or short circuiting techniques. Structural steels, such as A7, A36, common ship steels, piping, pressure vessel steels, and A515 Grades 55 to 70, are usually welded with this wire. Under the same conditions, weld beads are generally flatter and wider than those made with the ER70S-2 or the ER70S-3.

- **ER70S-5.** In addition to silicon and manganese, these wires contain aluminum as a deoxidizer. Because of the high aluminum content, they can be used for welding killed, semikilled, and rimmed steel with carbon dioxide shielding gas at high welding currents. Argon-oxygen and argon-carbon dioxide can also be used. Short-circuiting transfer should be avoided because of extreme pool viscosity. Rusty surfaces can be welded using this wire with little sacrifice in weld quality. Welding is restricted to the flat position.

- **ER70S-6.** One of the most commonly used of all the electrodes, this wire has high levels of manganese and silicon deoxidizers. See **Figure 8-1**. Like the ER70S-5, this wire yields quality welds on most carbon steels. This wire is ideal for use with material with medium to high levels of mill scale. It is applicable for both single and multipass welding applications with the use of carbon dioxide shielding gas and high welding currents. This electrode is also used with argon-oxygen mixtures containing 5% or more oxygen for high-speed welding. Since

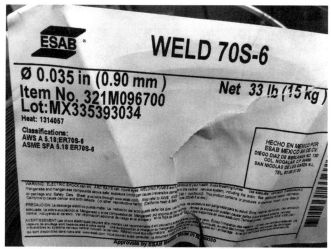

Goodheart-Willcox Publisher

Figure 8-1. Electrode type ER70S-6 is indicated on the reel tag. Always look for the AWS designation rather than the manufacturer's trade name for electrode designation.

ER70S-6 wire contains no aluminum, the short-circuit technique is possible with carbon dioxide or argon-carbon dioxide shielding gases. Like the ER70S-4, the weld pool is quite fluid.

- **ER70S-7.** This electrode wire has higher levels of manganese content than ER70S-6 but lower levels of silicon. The silicon levels are slightly higher than ER70S-3 electrodes. This wire can be used with argon/carbon dioxide gas mixtures. This chemical composition has intermediate hardness levels between an ER70S-3 and an ER70S-6 electrode.

- **ER70S-1B.** This wire contains silicon and manganese deoxidants plus molybdenum for increased strength. Welds can be made in all positions with argon-carbon dioxide and carbon dioxide shielding gases. Argon-oxygen is permitted for the flat position. Maximum mechanical properties are obtained with argon-oxygen and argon-carbon dioxide shielding gas mixtures. Slightly rusted surfaces can be welded with some sacrifice in weld quality. This wire is mostly used for welding low-alloy steels such as AISI 4130.

- **ER70S-G.** This *G* classification solid wire electrode does not have the AWS chemical composition, testing, or mechanical property requirements. Although these electrodes may meet or exceed the AWS requirements, they have not received the testing and documentation.

AWS A5.28 is the specification for low-alloy steel electrodes and rods for gas shielded arc welding. These electrode wires include the following:

- **ER80S-Ni1.** This wire is a high-silicon, high-manganese, low-alloy steel GMAW electrode containing 1% nickel. It also contains small amounts of chromium, molybdenum, and vanadium. These elements provide higher strength, impact properties, and corrosion resistance for weathering steels.

- **ER80S-D2.** This wire is a higher-silicon, higher-manganese, low-alloy electrode containing 0.50% of molybdenum. Molybdenum strengthens the weld metal and improves toughness properties.

The use of *metal cored electrode wire*, also known as *composite cored electrode wire*, has increased for production welding applications. Metal cored electrode wire is a tubular electrode that consists of a metal sheath and a core of various powdered materials, primarily iron. The core of metal cored wire contributes almost entirely to the deposited weld metal. The AWS A5.18 specification designates the composite design of this electrode with a *C*. The *R* is not included in the AWS designation because this electrode wire is not suitable as filler material for GTAW applications. The following are examples of this type of cored electrode wire:

- **E70C-6M.** This carbon steel electrode has high levels of silicon and manganese deoxidizers and is an excellent choice for welding on materials with high levels of mill scale. Due to its composite design, it is better suited than a solid wire electrode for these applications. This electrode is easy to use, has good weld pool fluidity, and has the ability to fill the weld pool at the weld toe, otherwise known as *wash in*. See **Figure 8-2**.
- **E90C-G.** This low-alloy electrode has a high-silicon, high-manganese deoxidizer necessary for welding over mill scale on high-strength steels. This electrode has high deposition rates with little slag or spatter.

Joint Preparation and Cleaning

Preparing joint edges with thermal cutting processes forms a heavy oxide scale on the surface. See **Figure 8-3**. Remove this oxide completely prior to welding to prevent porosity and dross within the weld. To a great extent, GMAW nullifies foreign material on the base metal. However, good welds cannot be made over oxide scale.

Remove scale and rough edges with a grinder. Remove foreign material next to the weld joint by sanding, **Figure 8-4**. When making multiple passes, clean the weld between each pass with a wire brush to remove oxides and foreign material. Where wire brushing does not remove the oxide scale, chipping

Figure 8-3. The oxide scale and small gouge indentations were formed on this bevel cut by the oxyacetylene cutting process. The scale must be completely removed prior to welding.

or grinding operations should remove it, **Figure 8-5**. Failure to remove these oxides may result in lack of fusion when the next pass is made.

Many materials received from the mill have a coating of oil to prevent rusting. Remove this oil with a cleaning solvent such as acetone before welding to prevent porosity. Do not weld near trichloroethylene vapor degreasers. The arc changes the vapor to poisonous phosgene gas.

Preheating, Interpass Temperature, and Postheating

Preheating, interpass temperatures, and postheating temperatures are recommended for optimum

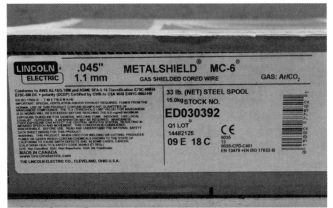

Figure 8-2. Metal cored electrode type E70C-6M is indicated on the reel tag. The AWS designation is indicated below the manufacturer's trade name.

Figure 8-4. The material has been sanded to a bright metal finish to prepare the part for welding. This type of sander is often called a *PG wheel*.

Figure 8-5. The oxide scale and silicon deposits formed during the welding process are removed prior to each weld by grinding, sanding, or wire brushing.

mechanical properties, crack resistance, and hardness control on various types of carbon and low-alloy steels. *Preheating* is the application of heat to the weld area to obtain a specific temperature prior to the start of welding. The preheated area should include the entire weld joint thickness and length.

Interpass temperature is the minimum/maximum temperature of the weld metal before the next pass in a multipass weld is made. Interpass temperature should always be maintained until the postheating operation is started.

Postheating refers to keeping the weldment at a specific temperature for a specific period of time after welding is completed. This operation should be started immediately after the welding is done. Upon completion of postheating, the weld is allowed to cool to room temperature in still air.

In general, carbon steels with less than 0.20% carbon and 1.0% manganese do not require any heating operation unless they are over approximately one inch thick, they are to be welded in severe restraint, or the metal temperature is less than 50°F (10°C).

Carbon steels over 0.020% carbon and 1.0% manganese are more sensitive to cracking. The increase in carbon and manganese, the degree of restraint, or an increase in weld thickness may require preheat, interpass temperature control, and postheat operations.

Carbon steels that contain carbon over 0.030% are often very crack sensitive and require heating and cooling temperatures and time rates. When establishing a weld procedure for these types of steels, consult the steel manufacturer and the wire manufacturer for these temperatures.

Low-alloy steels such as chrome-moly steels are often welded with the GMAW process. To prevent the formation of hard weld area grain structures, the amount of preheating is dictated by the joint thickness, restraint of the weldment, and the final postheating. Welding procedures should be established using a 200°F to 400°F (93°C to 204°C) range to determine if cracking occurs. Cracks indicate that the temperatures are too low and should be increased. Always allow any heated weldment to cool in still air or under a thermal blanket to slow down the cooling rate.

Quenched and tempered steels are a group of steels that have been heat treated for hardening and then tempered at a lower temperature to reduce the hardness and increase the ductility of the metal within specific tensile strength ranges. These steels are welded with the GMAW process. Contacting the manufacturer for the correct welding procedures required to maintain the desired mechanical values after welding is suggested.

Applying heat to the weld area is often accomplished with an oxyacetylene torch, **Figure 8-6**. A temperature-measuring instrument can be used to indicate the temperature. A temperature-indicating gun, crayons, or stickers, **Figure 8-7**, can also be

Figure 8-6. When heating with an oxyacetylene torch, do not use an excess acetylene flame adjustment. Always use a neutral flame.

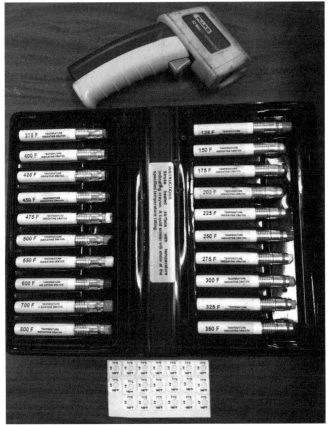

Goodheart-Willcox Publisher

Figure 8-7. Temperature guns, crayons, and other indicators are used to determine preheating temperatures prior to welding.

applied to the weld area. When the proper temperature is reached, the crayon melts. Do not place crayons on the weld joint because they can contaminate the joint.

Many weldments require another form of heating after the welding is completed in order to reduce the amount of stress in the weld area. *Stress relief* is usually accomplished in a furnace. The entire part is placed in the furnace for a period of time at a specific temperature. Due to the temperature tolerances and the extended time period for the operation, a welder using a manual oxyacetylene torch cannot do this operation.

Tooling

Tooling for GMAW is basically the same as for other welding processes. However, considerable spatter is generated during the GMAW short circuiting mode, which can cause problems within the tooling area. Copper, aluminum, or stainless steel tooling is generally used to minimize spatter pickup.

These materials also reduce the possibility of magnetic *arc blow* (the deflection of the intended arc pattern due to magnetism). In conjunction with the tooling materials, a liquid anti-spatter spray can be placed on the tooling areas where the spatter may fall to protect against spatter buildup. See **Figure 8-8**.

Considerations for designing tooling include the following:

- The proper weld joint alignment of the component parts.
- Heat control of the weld zone, if required.
- Allowance for shrinkage of the weld joint during welding.
- Providing atmosphere to prevent contamination of the weld crown and the weld penetration, if required.
- Assembly (loading) and disassembly (unloading) of the component parts.
- Positioning of the joint for welding.
- Accessibility for the welding gun and the welder to the weld joint.

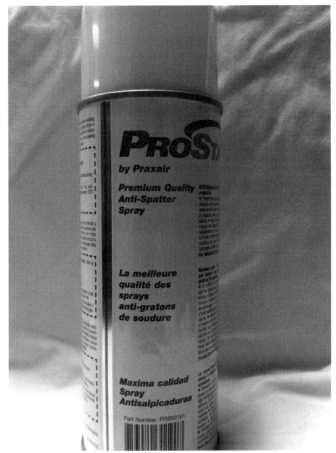

Goodheart-Willcox Publisher

Figure 8-8. An anti-spatter spray used to prevent spatter buildup on weldments.

- Tool inspection and maintenance program to maintain the integrity of the tool for the intended operation.

Welding Procedure

Establishing a welding procedure requires careful consideration of the following areas as they pertain to the specific operation:

1. Select the electrode wire and the shielding gas or gas combination to match the properties of the material.
2. Select the process mode.
3. Select the correct process parameters for the material thickness from setup charts.
4. Set the welding parameters on the power source and the wire feeder.
5. Install filler material, check drive rolls for proper type and size, and install the proper size contact tip and gas nozzle.
6. Select the proper shielding gas or gases and set the shielding gas flow rate (in cubic feet per hour, or cfh).
7. Set the shielding gas flow for the weld root side if required.
8. Establish the cleaning requirements.
9. Establish heating cycles, if required for the welding procedure.
10. Make a test weld on a piece of scrap material to determine the acceptability of the resulting weld. If tooling is designed for this operation, now is the time to test the tool and make any adjustments to be sure the tool is operating as designed. Adjust the welding equipment to fine-tune your welding parameters as you observe the welding operation.
    A. Establish tack weld procedures if required.
    B. Establish interpass temperatures to prevent cracking on multipass welds if required.
    C. Determine proper interpass cleaning procedures.
    D. Establish weld joint tolerances for fitup of the component parts.
    E. Use nondestructive tests and destructive tests to prove your final procedure and establish tolerances for inspection criteria.

11. During this period of weld testing, check your welding technique, focusing on the movement of the welding gun. Adjust gun movement as required.
12. Determine if postheating is required. Establish temperatures and periods of time if needed.
13. Establish the tolerances for each parameter and variable of your procedure. Each part of the procedure will not be exact; therefore, a tolerance must exist for each of these areas.

Plug and Slot Welding the Carbon Steels

Plug and slot welds are made with the same procedures used for fillet welds, except that the welder must manipulate the gun around the hole or slot. Thin-gauge materials can be welded in one pass. Heavier materials require several layers or passes. In these cases, always clean the completed weld with a wire brush before applying another pass or layer.

The major problem encountered in this type of weld is lack of fusion at the weld root. This is caused by an improper gun angle as the welder tries to manipulate the gun around the hole. To minimize this condition, change the gun angles as the weld progresses. Practice on sample weldments to develop the proper techniques, **Figure 8-9**.

Goodheart-Willcox Publisher

Figure 8-9. Completed plug and slot welds. After grinding or finishing, the weld is hard to detect.

Summary

- Carbon steels are commonly welded using the GMAW process. They are magnetic and melt at approximately 2500°F (1371°C).
- Carbon steel is classified based on the carbon content into low-, medium-, and high-carbon steel.
- Low-alloy, quenched and tempered, high-strength, and chrome-moly steel have additional elements added for specific mechanical properties and require special welding techniques.
- The two types of GMAW electrode wires are solid wire and composite metal-cored wire.
- A deoxidizer is an element that helps remove oxygen and nitrogen from the weld, reducing the occurrence of weld metal porosity. Materials with higher amounts of oxidation and mill scale require higher amounts of deoxidizers.
- Proper joint preparation is essential for a quality weld. Joint edges should be free of scale and contaminants.
- Some types of carbon steels require preheating, maintenance of interpass temperature, and postheating. These operations are recommended for optimum mechanical properties, crack resistance, and hardness control.
- Copper, aluminum, or stainless steel tooling minimizes spatter pickup in short circuiting GMAW. A liquid anti-spatter spray can be placed on tooling areas where the spatter may fall to protect against spatter buildup.
- An established weld procedure is essential for consistent quality welds. The weld procedure includes all variables required to produce a certain weld.

Review Questions

Answer the following questions using the information provided in this chapter.

1. Carbon steels have a maximum of 1.70% of _____.
2. Steel becomes _____ and more brittle as the percentage of carbon increases.
3. Hardening and tempering steel can improve the strength, _____ resistance, and corrosion resistance of the steel.
4. What does each letter and digit in ER70S-6 stand for?
5. Because of the high aluminum content of a(n) _____ electrode, it can be used for welding killed, semikilled, and rimmed steel with carbon dioxide shielding gas at high welding currents.
6. What does a *G* classification for a solid wire electrode mean?
7. How is each class of electrode wire defined?
8. Why are deoxidizers added to electrode wire?
9. List the five deoxidizers used in GMAW electrode wire.
10. Between each pass of a multipass weld, a wire brush should be used to remove _____ and foreign material.
11. What should be used to remove an oil coating on material to be welded?
12. Why should low-alloy steels be preheated?
13. Interpass temperature should be maintained until the _____ operation is started.
14. Copper, aluminum, or stainless steel tooling is generally used to minimize problems caused by _____ in the tooling area.
15. The major problem encountered in plug and slot welding is lack of _____ at the weld root.

Chapter 9

Gas Metal Arc Welding Procedures for Stainless Steels

Objectives

After studying this chapter, you will be able to:

❑ Explain the difference between various types of stainless steels.

❑ Select the appropriate electrode wire for specific welding conditions.

❑ Prepare a stainless steel joint for welding.

❑ Establish a GMAW welding procedure for stainless steel.

❑ Employ proper GMAW procedures for welding stainless steel.

❑ Understand the procedures for spot welding, plug welding, and slot welding stainless steels.

Technical Terms

austenite
austenitic stainless steels
carbide precipitation
duplex stainless steels
ferritic stainless steels
hot cracking
martensitic stainless steels
precipitation hardening stainless steels
stainless steels

Stainless Steels

There are many industrial applications for welding stainless steels. In order to meet the vast demands, a variety of stainless steels have evolved. *Stainless steels* are iron-based alloys that contain a minimum of 10.5% chromium. The chromium oxide film that forms on the surface of stainless steel is very thin and dense, yet it provides resistance to corrosion and prevents further oxidation.

There are five stainless steel families. Each contains many different types of stainless steels. The families are identified by their grain structure, which is formed during the manufacturing processes. The five stainless steel families are as follows:

- Austenitic stainless steels.
- Ferritic stainless steels.
- Martensitic stainless steels.
- Duplex stainless steels.
- Precipitation hardening stainless steels.

Austenitic Stainless Steels

Austenitic stainless steel is the most commonly used stainless steel. *Austenite* is a metallurgical term identifying a specific type of grain structure. The materials within the austenitic stainless steel group are identified by the American Iron and Steel Institute (AISI) as the 200 and 300 series. This group includes Types 302, 304, 310, 316, 321, and 347. See **Figure 9-1**.

The free machining grades (Types 303 and 303Se) contain sulfur that causes porosity, or holes in the weld, and *hot cracking*, which is the formation of shrinkage cracks during the solidification of the weld metal. Types 303 and 303Se are not recommended for welding applications.

Commercially Wrought Stainless Steel Identification (AISI)

				Composition, Percent[a]				
Type	C	Mn	Si	Cr	Ni	P	S	Others
302	0.15	2.00	1.00	17.0-19.0	8.0-10.0	0.045	0.03	
302B	0.15	2.00	2.0-3.0	17.0-19.0	8.0-10.0	0.045	0.03	
303	0.15	2.00	1.00	17.0-19.0	8.0-10.0	0.20	0.15 min	0-0.6 Mo
303Se	0.15	2.00	1.00	17.0-19.0	8.0-10.0	0.20	0.06	0.15 Se min
304	0.08	2.00	1.00	18.0-20.0	8.0-10.5	0.045	0.03	
304L	0.03	2.00	1.00	18.0-20.0	8.0-12.0	0.045	0.03	
305	0.12	2.00	1.00	17.0-19.0	10.5-13.0	0.045	0.03	
308	0.08	2.00	1.00	19.0-21.0	10.0-12.0	0.045	0.03	
309	0.20	2.00	1.00	22.0-24.0	12.0-15.0	0.045	0.03	
309S	0.08	2.00	1.00	22.0-24.0	12.0-15.0	0.045	0.03	
310	0.25	2.00	1.50	24.09-26.0	19.0-22.0	0.045	0.03	
310S	0.08	2.00	1.50	24.0-26.0	19.0-22.0	0.045	0.03	
314	0.25	2.00	1.5-3.0	23.0-26.0	19.0-22.0	0.045	0.03	
316	0.08	2.00	1.00	16.0-18.0	10.0-14.0	0.045	0.03	2.0-3.0 Mo
316L	0.03	2.00	1.00	16.0-18.0	10.0-14.0	0.045	0.03	2.0-3.0 Mo
317	0.08	2.00	1.00	18.0-20.0	11.0-15.0	0.045	0.03	3.0-4.0 Mo
317L	0.03	2.00	1.00	18.0-20.0	11.0-15.0	0.045	0.03	3.0-4.0 Mo
321	0.08	2.00	1.00	17.0-19.0	9.0-12.0	0.045	0.03	5 x %C Ti min
329	0.10	2.00	1.00	25.0-30.0	3.0-6.0	0.045	0.03	1.0-2.0 Mo
330	0.08	2.00	0.75-1.5	17.0-20.0	34.0-37.0	0.04	0.03	
347	0.08	2.00	1.00	17.0-19.0	9.0-13.0	0.045	0.03	C
348	0.08	2.00	1.00	17.0-19.0	9.0-13.0	0.045	0.03	0.2 Cu[b,c]
384	0.08	2.00	1.00	15.0-17.0	17.0-19.0	0.045	0.03	

a. Single values are maximum unless indicated otherwise. b. (Cb + Ta) min — 10 x %C. c. Ta — 0.10% max.

Goodheart-Willcox Publisher

Figure 9-1. The AISI identifies stainless steel by a numbering system and specifies the composition of each type. The column heading abbreviations represent carbon (C), manganese (Mn), silicon (Si), chromium (Cr), nickel (Ni), phosphorous (P), sulfur (S), molybdenum (Mo), selenium (Se), titanium (Ti), and copper (Cu).

The main elements of *austenitic stainless steels* are chromium and nickel. Chromium content is 16%–26% and nickel content is 8%–24%. These elements help the stainless steel resist corrosion.

The austenitic family is welded more than any of the other steel series. It has good corrosion resistance, excellent strength at both low and high temperatures, and a high degree of toughness. Austenitic stainless steel cannot be hardened by heat treatment and is nonmagnetic.

Ferritic Stainless Steels

The *ferritic stainless steels* contain 10.5%–30% chromium and up to 1.20% carbon, along with small amounts of manganese and nickel. This group includes Types 405, 409, 430, 442, and 446. Because they are ferritic at all temperatures and do not transform to austenite, they cannot be hardened by heat treatment. Ferritic stainless steels are magnetic.

Martensitic Stainless Steels

The *martensitic stainless steels* contain 11%–18% chromium and up to 0.20% carbon. This group includes Types 403, 410, 414, 416, 420, 422, 431, and 440.

These steels transform to austenite when heated and then can be transformed to martensite with cooling. They have a tendency toward weld cracking during cooling when the hard, brittle martensite is formed. The martensitic stainless steels are magnetic.

Duplex Stainless Steels

The *duplex stainless steels* are part ferritic, part austenitic. They solidify as 100% ferrite, but half of the ferrite transforms into austenite during cooling. This type has a higher chromium and lower nickel content than austenitic grades. Duplex steels are magnetic. The most common grade is Type 2205.

Precipitation Hardening Stainless Steels

The *precipitation hardening (PH) stainless steels* are further divided into martensitic, semi-austenitic, and austenitic. Precipitation hardening, also called *age hardening*, is a heat treatment that increases the yield strength of malleable materials. PH stainless steels combine high strength and hardness with corrosion resistance, making them superior to martensitic stainless steels. Depending on the grade and intended use, ductility is improved by temper procedures. The various procedures for quenching and tempering these stainless steels result in various hardness levels necessary for the preferred applications. Examples from this group are 17-4PH, 17-7PH, 15-5PH, PH 15-7 Mo, and AM 350. Electrodes for these materials should match the alloy base metal.

Stainless Steel Electrode Wire

Various electrodes are used to join stainless steel alloys. The electrodes should be compatible with the base material. See **Figure 9-2.** They should provide crack-resistant deposits, which are equal to or better than the base material in terms of soundness, corrosion resistance, strength, and toughness. Since weld metal with a composition equal to the base metal does not usually match the properties or performance of wrought base metal, it is common to enrich weld metal alloy content. Always use low-carbon (LC) or extra-low-carbon (ELC) stainless steel electrode wire when it is specified.

Protect the electrode wire material from contamination at all times. When the wire is installed on a machine, a spool cover should always completely enclose the material. Store wire removed from the machine in a container to prevent the exposure to foreign material.

The electrodes commonly used for welding the austenitic stainless steels are defined by the American Welding Society AWS A5.9 *Specification for Bare Stainless Steel Welding Electrodes and Rods.* The following types of electrode wire can be used for stainless steel welding:

- **ER308.** This electrode wire is commonly used to weld type 304 stainless steel or any base material containing approximately 19% chromium and 9% nickel.

- **ER308L.** Wires of this type can be used for the same materials as ER308. The chromium and nickel contents are identical to ER308, but the lower carbon content reduces any possibility of carbide precipitation and the intergranular corrosion that can occur.
- **ER308L Si.** This electrode wire has chemistry similar to ER308 wire. See **Figure 9-3.** However, a higher silicon level improves the wetting characteristics of the weld metal, particularly when argon-1% oxygen shielding gas is used. If the dilution of the base metal is extensive, high silicon content can cause greater crack sensitivity than a lower silicon content. This results from the weld being fully austenite or low ferrite.
- **ER309.** This electrode wire is used to weld type 309 and 309S stainless steel. It can be used to weld type 304 stainless steel where severe

Goodheart-Willcox Publisher

Figure 9-3. This ER308L Si type electrode has chemistry similar to the ER308 wire, with a higher silicon level that improves the wetting characteristics of the weld bead.

Weld Filler Material										
	308	308L	309	310	316	316L	317	317L	330	347
Base Material	301 302 304 304N 308	304L 310S	309 309S 384	310	316 316N	316L	317	317L	330	321 347 348

Goodheart-Willcox Publisher

Figure 9-2. The welding wire listed at the top of each column is used to weld the base materials listed below it.

corrosion conditions will be encountered and for joining mild steel to type 304.

- **ER316.** This electrode wire is used to weld types 316 and 319 stainless steels. The addition of molybdenum makes this wire electrode useful for high-temperature service where creep resistance is desired. Creep failure is caused by continuous strain over long periods of time at loads below the yield point of the material. Creep happens only at high temperatures of 900°F (482°C) or higher for stainless steels.
- **ER316L.** This electrode wire, due to the lower carbon content, is less susceptible to intergranular corrosion caused by carbide precipitation when used in place of ER316.
- **ER347.** This electrode wire is much less subject to intergranular corrosion from carbide precipitation because tantalum and/or columbium are added as stabilizers. This wire is used for welding base materials with similar chemistry and where high-temperature strength is required.
- **ER410.** This electrode wire is used for the martensitic stainless steels 402, 410, 414, and 420. When the wire is required to match the carbon content in type 420, ER420 should be used.
- **ER409** and **ER430.** These electrodes should be used to match the ferritic stainless steels or exceed the chromium levels of the base material.

Joint Preparation and Cleaning

Standard weld joint designs can be used for GMAW stainless steel applications. In some cases involving thin metals, backing bars are used to prevent oxidation of the weld penetration. In some cases where the two materials are not compatible, one of the materials must be cladded for a compatible weld. See **Figure 9-4.**

Dirty weld joints lead to porosity and carbon pickup. Therefore, remove all dirt, grease, crayon marks, and other foreign matter before assembling the parts to be joined.

When wire brushing is required prior to or after welding, always use a stainless steel wire brush that has not been used on other materials, **Figure 9-5**. A carbon steel brush could induce contamination, causing surface rusting. The same requirement applies to grinding disks. Only use disks that are designated for the stainless steel application.

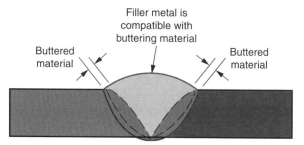

Goodheart-Willcox Publisher

Figure 9-4. A V-groove joint of two different materials. When materials to be joined are not compatible, one or both materials can be cladded (buttered) to make an acceptable weld with a common filler material.

Tooling

Nonmagnetic tooling must always be used for welding in order to prevent arc blow. Magnetic tooling interferes with the flow of metal from the electrode and causes defects in the completed weld. Since stainless steel retains heat, the tooling must remove the heat to reduce warpage, avoid distortion, and prevent carbide precipitation of the weld and heat-affected zone. See **Figure 9-6**. *Carbide precipitation* is the movement of chromium out of the grain into the grain boundary. The depletion of chromium decreases the amount of corrosion protection in the material.

Copper is nonmagnetic and is a good backing material because it removes heat from the stainless steel very rapidly. The tooling can be machined to provide space for weld penetration through the back side of the joint. Shielding gas can be used to prevent oxidation of the weld root side. A backing bar

Goodheart-Willcox Publisher

Figure 9-5. Use stainless steel wire brushes dedicated for cleaning the stainless steel base metal and weld in order to avoid contamination.

Goodheart-Willcox Publisher

Figure 9-6. Warpage and distortion of the stainless steel after just a couple of weld beads.

designed for admitting the shielding gas to the root or underside of the weld is shown in **Figure 9-7**. Always protect the molten metal on the root side of the weld to prevent the absorption of oxygen and nitrogen during the solidification and cooling period.

The root side of the weld joint can also be protected with a flux designed for this particular application. Mix the flux powder with acetone or alcohol and make it into a paste. Apply the paste to the root side of the joint with a brush. The alcohol or acetone evaporates, leaving the flux adhering to the joint area. The paste melts and flows over the molten metal during the welding operation to prevent oxidation. After welding, remove the remaining flux by brushing the part in hot water or with a commercial acid cleaner.

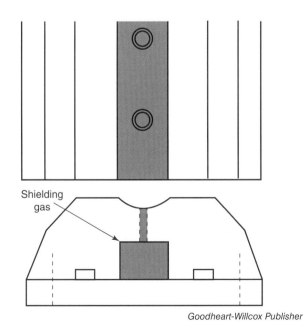

Goodheart-Willcox Publisher

Figure 9-7. Backing bar designed for admitting gas to the penetration side of the weld joint.

Welding Procedures

The GMAW process with solid or metal-cored electrode wire is the best choice for stainless steel welding operations that involve long joints or many parts. The following standard modes of metal transfer are used for gas metal arc welding of stainless steel:

- **Short circuiting transfer.** See **Figure 9-8** and **Figure 9-9**.
- **Spray transfer.** See **Figure 9-10**. With the increased voltage required for spray transfer, the amount of heat input increases also.

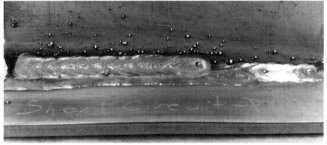

Goodheart-Willcox Publisher

Figure 9-8. This weld was made with the short circuiting GMAW process in the backhand (pull) direction. Note the amount of spatter.

Goodheart-Willcox Publisher

Figure 9-9. This piece was welded with the short circuiting GMAW process in the forehand (push) direction. Note the amount of spatter.

Goodheart-Willcox Publisher

Figure 9-10. The spray transfer mode produces a clean weld, but the excessive voltage also increases undesirable heat input on thin material.

- **Pulse spray transfer.** For a deep penetration weld with minimal spatter and heat input, the pulse spray transfer mode produces a high quality weld. See **Figure 9-11**.

Globular transfer is not generally used for stainless steel. This mode is difficult to use with out-of-position welding.

The welding procedure for stainless steels must be set up to avoid contamination of the molten weld pool. The gas nozzle must be of sufficient size to cover the weld, and the electrode extension during welding must be held to procedure tolerances. Avoid welding in drafty areas where gas coverage of the molten pool may be lost.

The shielding gas recommended for short circuiting transfer welding of stainless steels is a tri-mix blend of helium, argon, and carbon dioxide, **Figure 9-12**. Check gas nozzles often to clear spatter from inside the nozzle that could interfere with the proper gas flow.

The electrode wire diameters generally used for GMAW of stainless steels are between 0.030″ and 3/32″ (0.8 mm and 2.4 mm). Diameters of 0.035″ and 0.045″ (0.9 mm and 1.1 mm) are most commonly used for this process.

Two problems associated with the welding of stainless steels are the sensitization of the weld's heat-affected zone and the hot cracking of the weld metal. Sensitization is caused by chromium carbide formation and precipitation in the heat-affected zone adjacent to the weld. This results in the breakdown of the chromium and reduced corrosion resistance. Welds should be made without preheat and with minimal heat input. Rapid cooling after welding can reduce the amount of carbide precipitation.

General welding parameters with the required shielding gas are listed in the *Reference Section* for both short circuiting transfer and spray transfer modes. The pulsed spray transfer mode is not referenced, since different welding power sources use different pulse

Figure 9-12. A tri-gas mixture of helium, carbon dioxide, and argon is ideal for most stainless steel welding.

frequencies. Refer to the welding machine manual for setup parameters for this type of machine.

Tips for Welding of Stainless Steel

Welding procedures must be set up prior to welding to prevent contamination of the weld. Review the following keys to successful welding of stainless steel prior to starting a weld:

- When installing a roll of wire on a machine that has been used for other materials, clean the entire wire feed system, including the drive rollers and conduit liners.
- Select the proper electrode wire for the base stainless steel type.
- Always hold the gas nozzle as close to the work as possible.
- Use forehand welding in the pulsed spray transfer and spray transfer modes. The forehand welding technique forms a flatter weld crown on butt welds and provides better visibility of the weld as it is being made.
- Use a slight back and forth whipping motion for short circuiting transfer. Backhand welding techniques are usually easier for making fillet welds and result in a neater weld profile.
- Stringer beads reduce the overall heat input and the possibility of cracking.
- When making a multipass weld, always wire brush the previous pass with a stainless steel brush. Do not use stainless steel brushes that have been used on carbon steel material because they can cause contamination.

Figure 9-11. The pulsed spray transfer mode produces high quality welds without excessive heat input.

- Do not leave craters at the end of the weld. Use starting and run-off tabs when automatic welding and weld back over the end in the semiautomatic mode. Craters are prone to cracking.
- When welding magnetic stainless steels (ferritic and martensitic types) to the relatively nonmagnetic types (austenitic), use a single bevel joint, weld in the short circuiting transfer mode, and keep the electrode centered over the nonbeveled edge of the weld joint.
- Carefully control the amount of heat in the heat-affected zone to reduce the levels of carbide precipitation and reduce hot cracking problems.

Spot Welding Stainless Steels

Select a welding wire type from the table shown in **Figure 9-2** (or as specified) and set up the welding equipment. Consider the type of process to be used, the wire diameter, and the appropriate shielding gas.

The austenitic stainless steels are welded with the same equipment and procedures used for the carbon steels (described in Chapter 8). Whenever possible, use a backing gas or a commercial flux to prevent oxidation of the weld penetration when spot welding corner joints and lap joints where full penetration is required.

A major problem encountered in welding thin sections is distortion during the welding and cooling of each spot weld. Tooling may be required to contain warpage and maintain weld joint alignment.

Lap joints must be cleaned if an oxide scale is present. Clean with a stainless steel wire brush or use a commercial cleaner if fusion between the plates is to be complete.

Plug and Slot Welding Stainless Steels

Use the same base metal and filler metal combinations as previously described, and set up the process with the proper gas for the welding mode. Welding techniques are the same as those used for welding the carbon steels.

The major problem encountered in plug and slot welding is the condition of the holes or slots. Make sure that the hole or slot is free from burrs, slag (if plasma cut), or cutting oils used for drilling. Practice these types of welds in order to develop the correct technique needed to produce good fusion.

Summary

- The five types of stainless steels are austenitic, ferritic, martensitic, duplex, and precipitation hardening. The types are based on grain structure developed during manufacturing.
- The main element that distinguishes stainless steel is chromium. These iron-based alloys have a minimum of 10.5% chromium.
- The electrode wire should be determined based on the type of stainless steel and should match the base metal as closely as possible.
- The use of preheat before welding and high interpass temperatures would overheat the stainless steel and result in carbide precipitation and hot cracking, which are undesirable.
- Proper precautions should be taken to avoid contamination of the stainless steel base metal and electrode wires. The use of a stainless steel wire brush and designated grinding disks are required during preparation and cleaning of the base metal.
- It is important to determine and follow proper welding procedures in order to produce a quality weld.

Review Questions

Answer the following questions using the information provided in this chapter.

1. Stainless steels contain a minimum of 10.5% _____.
2. *True or False?* Austenitic stainless steel is the most commonly used stainless steel.
3. Chromium and nickel are the main elements of _____ stainless steels.
4. Types 420 and 422 are examples of _____ stainless steels.
5. Duplex steels are part _____, part austenitic.
6. Electrodes used to join stainless steel alloys should be compatible with the _____.
7. What is the difference between ER308 and ER308L electrode wires?
8. *True or False?* A carbon steel brush should be used for wire brushing prior to and after welding.
9. In order to prevent arc blow, _____ tooling must be used for welding.
10. Why is copper a good backing material?
11. What type of shielding gas is recommended for short circuiting transfer welding of stainless steels?
12. *True or False?* Preheating is recommended before stainless steels are welded.
13. The _____ welding technique should be used in the pulsed spray transfer and spray transfer modes.
14. List two reasons why the amount of heat in the heat-affected zone should be carefully controlled.
15. When plug or slot welding stainless steels, make sure the hole or slot is free from _____, slag, or cutting oils.

Chapter 10

Gas Metal Arc Welding Procedures for Aluminum

Objectives

After studying this chapter, you will be able to:
- ❏ Describe the characteristics of aluminum.
- ❏ Identify and describe alloying elements for aluminum.
- ❏ Select the appropriate electrode wire for different welding conditions.
- ❏ Prepare an aluminum joint for welding.
- ❏ Weld aluminum using proper techniques.
- ❏ Understand the problem of burnback and how to avoid it.

Technical Terms

burnback
nonferrous metal
oxide
run-off tab
temper designations

Aluminum

Aluminum is a *nonferrous metal* (one that contains no iron) that can be readily welded with the GMAW process. Characteristics of aluminum include the following:
- High thermal conductivity.
- Good electrical conductivity.
- High ductility at subzero temperatures.
- Light weight.
- High corrosion resistance.
- Nonsparking.
- Nontoxic.
- No color change when heated.

Pure aluminum melts at approximately 1200°F (647°C). Aluminum alloys have a melting point between approximately 900°F and 1200°F (482°C and 647°C). An *oxide* film (material formed through the chemical combination of the metal and oxygen) forms on the surface of aluminum during manufacturing. This film provides good corrosion resistance; however, it must be removed prior to welding in order to produce quality welds.

Metallurgy of Aluminum

Aluminum is the most abundant metallic element in the Earth's crust. Bauxite is the name of the rock that contains the aluminum oxides and the main source for aluminum. Aluminum is not found in nature in the metallic form.

The aluminum family is made up of pure aluminum and various alloying elements that provide specific characteristics to the base material. There are three major differences between welding steel and welding aluminum—some aluminum alloys cannot be arc welded, most aluminum welds are not as strong as the base material, and only some aluminum alloys are heat-treatable.

Alloying Elements

Alloys are assigned specific designations for identification based on the alloying elements added to the aluminum. See **Figure 10-1**.

- **Pure aluminum (1xxx series)** contains no additional alloying materials. Pure aluminum is weldable with 1100 and 4043 electrode wire.
- **Copper (2xxx series)** adds a higher strength to the aluminum. These alloys are more sensitive to hot cracking. Most of these alloys should not be welded. The exceptions are alloys 2014, 2219, and 2519, which are weldable with 4043 or 2319 electrodes.
- **Manganese (3xxx series)** creates moderate strength aluminum for sheet applications. This alloy is excellent for welding with either 4043 or 5356 electrode wire and is not susceptible to hot cracking.
- **Silicon (4xxx series)** reduces the melting point and improves the fluidity of the material. The main use for this series is electrode wire, such as 4043.
- **Magnesium (5xxx series)** improves weldability, adds structural strength, and is not prone to hot cracking. This series has the highest strength of the nonheat-treatable alloys.
- **Silicon and magnesium (6xxx series)** produces a medium strength material. This alloy is somewhat prone to hot cracking, but the cracking can be reduced with the proper joint and electrode

Wrought Aluminum Alloy Groups	
Alloy Group	**Series Designation**
Aluminum, 99% minimum purity	1XXX
Aluminum-copper	2XXX
Aluminum-manganese	3XXX
Aluminum-silicon	4XXX
Aluminum-magnesium	5XXX
Aluminum-magnesium-silicon	6XXX
Aluminum-zinc	7000
Other	8000

Goodheart-Willcox Publisher

Figure 10-1. Designations for wrought aluminum alloy groups.

selection. This aluminum series should be welded with either 4xxx or 5xxx electrode wire, with the 4043 most commonly used.

- **Zinc (7xxx series)** is added along with magnesium and copper to produce the highest strength heat-treatable series of aluminum. This alloying element is primarily used in the aircraft and bicycle industry due to the high strength to weight characteristics. Grades 7005 and 7039 are weldable with 5356 electrodes.
- **Other (8xxx series)** includes aluminum alloyed with elements such as lithium. Though most of this series is not easily weldable, the 4000 electrodes are the primary filler to use.

There are also many secondary alloying elements, including chromium, iron, zirconium, vanadium, bismuth, nickel, and titanium. These elements are alloyed with aluminum to improve strength, corrosion resistance, and heat treatability.

Nonheat-Treatable Wrought Alloys

The aluminum materials in this family include the 1000, 3000, 5000, and some of the 4000 series. They contain specific alloying elements for strength; however, they do not respond to heat treatment for higher strength. Some of these materials harden when cold-worked during the manufacturing process. Heat from the welding process may remove some of this strength in the heat-affected zone during welding.

Heat-Treatable Wrought Alloys

The heat-treatable alloys include materials in the 2000, 6000, and 7000 series. Some 4000 alloys are also heat-treatable. These materials are made in various heat-treated conditions, which are identified by *temper designations*. The basic aluminum heat treatment temper designations are: F (fabricated), O (annealed, or the lowest strength), H (strain hardened), and T (solution heat-treated). Additional numbers are added to the designation to describe the treatment process. See **Figure 10-2**. These materials have higher strength in the annealed condition than the nonheat-treatable alloys, but the weld zone may soften during welding. Heat-treatable alloys may require further heat treatment to regain their preweld condition.

Casting Alloys

Aluminum castings are manufactured for pump housings, engine frames, motor bases, and similar uses. These alloys are identified by a different numerical system, **Figure 10-3**. Aluminum castings are produced by sand mold, permanent mold, or

Basic Temper Designations for Aluminum Alloys	
Designation	**Condition**
F	As fabricated.
O	Annealed.
H1	Strain-hardened only.
H2	Strain-hardened and partially annealed.
H3	Strain-hardened and thermally annealed.
W	Solution heat-treated.
T1	Cooled from elevated-temperature shaping process and naturally aged.
T2	Cooled from elevated-temperature shaping operation, cold-worked, and naturally aged.
T3	Solution heat-treated and then artificially aged.
T4	Solution heat-treated and naturally aged.
T5	Cooled from elevated-temperature shaping process, and then artificially aged.
T6	Solution heat-treated, then artificially aged.
T7	Solution heat-treated and stabilized.
T8	Solution heat-treated, cold-worked, then artificially aged.
T9	Solution heat-treated, artificially aged, and then cold-worked.
T10	Cooled from an elevated-temperature shaping process, cold-worked, and then artificially aged.

Goodheart-Willcox Publisher

Figure 10-2. Temper designations for aluminum alloys.

Cast Aluminum Alloy Groups	
Alloy Group	**Series Designation**
Aluminum, 99% purity	1XX.X
Aluminum-copper	2XX.X
Aluminum-silicon-copper or aluminum-silicon-magnesium	3XX.X
Aluminum-silicon	4XX.X
Aluminum-magnesium	5XX.X
Aluminum-zinc	7XX.X
Aluminum-tin	8XX.X
Other alloy systems	9XX.X

Goodheart-Willcox Publisher

Figure 10-3. Designations for cast aluminum alloy groups.

die casting methods. These castings may be either nonheat-treatable or heat-treatable, with temper designations the same as those for wrought aluminum.

Electrodes for Aluminum Alloys

Selection of the proper electrode wire for welding aluminum is essential in order to produce a strong, quality weld that can be used in the same applications as the base material. The common electrode filler materials used for welding aluminum are defined by specification AWS A5.10 *Welding Consumables—Wire Electrodes, Wires and Rods for Welding of Aluminum and Aluminum-Alloys—Classification.*

Figure 10-4 lists electrode wires for welding various alloys, while electrodes for welding aluminum with specific properties are shown in **Figure 10-5**. The two most common electrodes for welding aluminum are ER4043 and ER5356. See **Figure 10-6**.

When selecting weld wire diameter, consider the type of power source, welding mode, type of wire feeder equipment, type of welding gun, joint design, and position of the weld. Also consider the operator's ability to manipulate the welding gun under adverse conditions if preheat is used. Push-pull gun or spool gun systems are generally used for welding aluminum due to the soft electrode wire. If a push wire feeder and gun are used, the gun cable should be as straight as possible to avoid feed problems.

Electrode materials should be stored and maintained under conditions that prevent oxidation of the material. Oxide scale on the wire causes defects in the weld, such as gas pockets and porosity.

Joint Design

Fillet welds do not generally present a problem when aluminum is welded because the weld shrinkage is not serious. In a butt weld, however, the center of the groove weld shrinks last, resulting in centerline cracking. This problem becomes more serious as the filler material combines with and dilutes into the base material at high temperatures. The best way to reduce dilution of the electrode wire and the base material is to use a V-groove or U-groove weld whenever possible. Do not use the square-groove butt weld design—too much dilution will take place and the weld may be crack-sensitive.

Joint Preparation and Cleaning

Preparation of weld joint edges by plasma arc cutting or removal of cracks with carbon arc gouging results in a heavy aluminum oxide scale on the base metal surface. Aluminum cannot be cut using the

Base Metal	6070	6061, 6063 6101, 6151 6201, 6951	5456	5454	5154a 5254a	5086	5083	5052 5652a	5005 5050	3004 Alc. 3004	2219	2014 2024	1100 3003 Alc. 3003	1060 EC
1060, EC	ER4043h	ER4043h	ER5356c	ER4043e,h	ER4043e,h	ER5356c	ER5356c	ER4043i	ER1100c	ER4043	ER4145	ER4145	ER1100c	ER1100
1100, 3003 Alclad 3003	ER4043h	ER4043h	ER5356c	ER4043e,h	ER4043e,h	ER5356c	ER5356c	ER4043e,h	ER4043e	ER4043e	ER4145	ER4145	ER1100c	
2014, 2024	ER4145	ER4145									ER4145a	ER4145a		
2219	ER4043f,h	ER4043f,h	ER4043	ER4043h	ER4043h	ER4043i	ER4043	ER4043i	ER4043	ER4043	ER2319c,f,h			
3004 Alclad 3004	ER4043e	ER4043b	ER5356e	ER5654b	ER5654b	ER5356e	ER5356e	ER4043e,h	ER4043e	ER4043e				
5005, 5050	ER4043e	ER4043b	ER5356e	ER5654b	ER5654b	ER5356e	ER5356e	ER4043e,h	ER4043d					
5062, 5652 a	ER5356b,c	ER5356b,c	ER5356b	ER5654b	ER5654b	ER5356e	ER5356e	ER5654a,b,c						
5083	ER5356e	ER5356e	ER5183e	ER5356e	ER5356e	ER5356e	ER5183							
5086	ER5386e	ER5356e	ER5356e	ER5356e	ER5356b	ER5356e								
5154, 5254 a	ER5356b,c	ER5356b,c	ER5356b	ER5654b	ER5654a,b									
5454	ER5356b,c	ER5356b,c	ER5356b	ER5554a,b										
5456	ER5356e	ER5356e	ER5556e											
6061, 6063 6101, 6201, 6151, 6951	ER4043b,h	ER4043b,h												
6070	ER4043e,h													

Where no filler metal is listed, base metal combination is not recommended for welding.

a. Base metal alloys 5652 and 5254 are used for hydrogen peroxide service. ER5654 filler metal is used for welding both alloys for low-temperature service (150°F and below).
b. ER5183, ER5356, ER5554, ER5556, and ER5654 may be used.
c. ER4043 may be used.
d. Filler metal with same analysis as base metal is sometimes used.
e. ER5183, ER5356, or ER5556 may be used.
f. ER4145 may be used.
g. ER2319 may be used.
h. ER4047 may be used.
i. ER1100 may be used.

Goodheart-Willcox Publisher

Figure 10-4. Wrought aluminum alloy filler material selection.

Base Metal	Filler Alloys[1]		Base Metal	Filler Alloys[1]	
	Preferred for Maximum As-welded Tensile Strength	Alternate Filler Alloys for Maximum Elongation		Preferred for Maximum As-welded Tensile Strength	Alternate Filler Alloys for Maximum Elongation
EC	1100	EC/1260	5086	5183	5183
1100	1100/4043	1100/4043	5154	5356	5183/5356
2014	4145	4043/2319 [3]	5357	5554	5356
2024	4145	4043/2319 [3]	5454	5554	5356
2219	2319	—	5456	5556	5183
3003	5183	1100/4043	6061	4043/5183	5356 [2]
3004	5554	5183/4043	6063	4043/5183	5183 [2]
5005	5183/4043	5183/4043	7039	5039	5183
5050	5356	5183/4043	7075	5183	—
5052	5356/5183	5183/4043	7079	5183	—
5083	5183	5183	7178	5183	—

The above table shows recommended choices of filler alloys for welds requiring maximum mechanical properties. For all special services of welded aluminum, inquiry should be made of your supplier.
1. Data shown are for "0" temper.
2. When making welded joints in 6061 or 6063 electrical conductor in which maximum conductivity is desired, use 4043 filler metal.

However, if strength and conductivity both are required, 5356 filler may be used and the weld reinforcement increased in size to compensate for the lower conductivity of the 5356 filler metal.
3. Low ductility of weldment is not appreciably affected by filler used. Plate weldments in these base metal alloys generally have lower elongations than those of other alloys listed in this table.

Goodheart-Willcox Publisher

Figure 10-5. Filler materials used to obtain specific properties in completed welds.

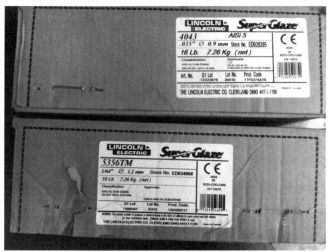

Figure 10-6. ER4043 and ER5356 are the two most common aluminum electrode types.

oxyacetylene process. All surfaces to be welded must be cleaned to bare metal prior to welding. Any oxidized metal produced by cutting or gouging remains in the weld, causing severe porosity. The heavy oxidation must be removed by machining or sanding. Several types of tungsten carbide and tool steel rotary files, sanding disks, and specially designed grinding disks can be used for this purpose. See **Figure 10-7**. Never use regular grinding disks to remove the oxide film. Use only a grinding wheel specifically designed for aluminum to clean the oxide film from aluminum.

Joint edges prepared by the shearing process should be sharp, without showing tearing or ridges. Any tearing or ridges can trap dirt or oil, resulting in faulty welds. Straight edges with these types of imperfections should be filed to clean metal.

Figure 10-7. Carbide rotary files, sanding disks, and aluminum designated grinding wheels are used to remove dross left from plasma arc cutting.

Cleaning of joints in preparation for welding can be done by chemical or mechanical means. Chemical methods include commercial degreasers, commercial cleaning compounds, and chemical baths. Use caution when working with chemicals. Always wear proper PPE to protect eyes, skin, and breathing. Use the proper disposal methods recommended by the chemical manufacturer.

Mechanical cleaning methods include filing or scraping, sanding with abrasive pads or abrasive wheels, and stainless steel brushes. Do not use carbon steel brushes or stainless steel brushes that have been used on carbon steel. It is a good practice to mark the brush "For Aluminum Only." See **Figure 10-8**.

After cleaning is completed, but before welding, wash the entire weld joint with alcohol or acetone. Allow the metal to dry completely before welding. Do not weld on a wet surface. Clean the weld joint and the immediate weld area immediately before welding. If welding must be delayed, cover the weld area with plain brown wrapping paper and seal it with tape to keep the joint clean.

Tooling

Tooling for backing is needed when full penetration groove welds are made without an aluminum backing strip. The molten weld penetration must be prevented from dropping through into air, since this causes excessive oxidation and impedes proper weld pool flow. Wherever possible, use a solid aluminum or copper backing bar with a groove for the drop-through. It will prevent oxidation and establish a base for the weld root.

When high-quality welds are required, purging of the root or back of the weld is necessary. The backing tooling shown in **Figure 10-9** can be used for this purpose. This tool has a groove machined into its face for the weld drop-through, and holes are drilled into a manifold in its base. The manifold is connected to a supply of argon or helium gas. The gas is admitted

Figure 10-8. A stainless steel wire brush marked for aluminum only use.

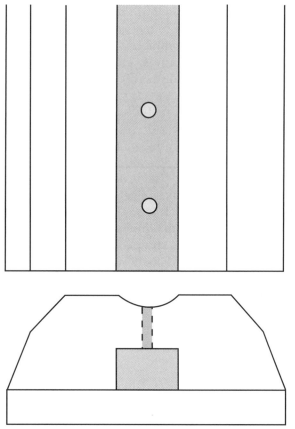

Goodheart-Willcox Publisher

Figure 10-9. When argon gas is admitted to the lower section of the tool, it flows through the holes to the grooved area. The purge gas shields the penetration during welding, prevents the formation of oxides, and assists in forming the root bead.

during welding to protect the weld drop-through from contamination. Copper or stainless steel hold-down bars are often used to secure the joint to the fixture.

Welding Procedures

The short circuiting, spray transfer, and pulsed spray transfer modes can be used for welding aluminum. The short circuiting mode, **Figure 10-10**, is difficult to use unless the procedure is tightly controlled for the electrode stickout requirement. This generally requires an automatic welding operation. When used in combination with the proper shielding gas, the spray transfer and pulsed spray transfer modes offer lower overall current levels that can be used in all thicknesses and positions. See **Figure 10-11** and **Figure 10-12**.

Argon is the main gas used for welding of aluminum, since it provides good arc stability with the least amount of spatter. However, the addition

Run-off
tab

Goodheart-Willcox Publisher

Figure 10-10. This aluminum weld was performed in the short circuiting mode on a test plate. Note the run-off tab.

Goodheart-Willcox Publisher

Figure 10-11. This aluminum multipass fillet weld was completed using the spray transfer mode.

Goodheart-Willcox Publisher

Figure 10-12. This aluminum multipass fillet weld was made in the pulsed spray transfer mode.

of helium often yields a hotter molten weld pool with better penetration on thicker material and less porosity.

The shielding gas flow rates with a nozzle of 1/2″ to 5/8″ diameter should be approximately 25 to 40 cfh (cubic feet per hour). Gas flow rates at the higher end of the range should always be used with helium mixes. Protect the arc from drafts and breezes. Tables in the *Reference Section* of this book are useful for the initial setup of the voltage and wire feed speed of an aluminum welding procedure based on the transfer mode, electrode wire diameter, and base material thickness.

Preheating is a common practice in welding of aluminum. Preheating refers to heating the base metal prior to welding, which reduces the amount of welding heat needed to achieve penetration and increases welding speed. Oxyacetylene heating can be used, but the upper limit of heating should not exceed 300°F (149°C). See **Figure 10-13**. Temperature-indicating guns, crayons, or special indicating stickers can be used to check this temperature range. Refer to **Figure 8-7**. Ensure that the temperature-indicating material does not come into contact with the weld area.

When preheating, always remember that the welding procedure will require less welding current for the same amount of penetration, and the weld pool will be larger. As the material heats up during welding, the amperage may be reduced slightly.

When making a multipass weld, always clean the completed weld with a stainless steel brush before applying the next pass. See **Figure 10-14**.

For short circuiting transfer or spray transfer welding on fillet or butt welds, the forehand welding technique produces good cleaning action of the weld joint. See **Figure 10-15**. The forehand or push direction cleans and etches the aluminum as it is welded.

Avoid stopping the weld at the end of a joint and leaving a crater, **Figure 10-16**. Aluminum is prone to cracking when the weld cross-section is thin or a concave shape. Using the same gun angle, always move back onto the end of the weld before stopping in order to fill the crater. Where this cannot be done, use a *run-off tab*, which is a short piece of base metal tack welded to the end of the joint, and then run the weld over the end before stopping. After the weld is completed, the run-off tab is removed. See **Figure 10-17**.

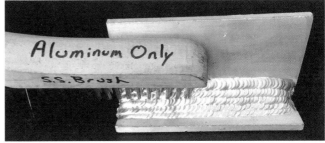

Goodheart-Willcox Publisher

Figure 10-14. When multipass welding on aluminum, use a stainless steel wire brush to clean the weld joint after each pass.

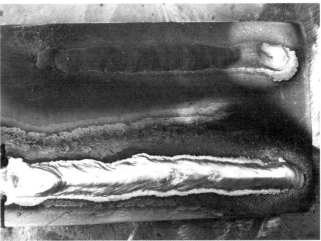

Goodheart-Willcox Publisher

Figure 10-15. The dark-colored weld was made with the backhand welding technique (note the soot on and around the bead). The light-colored section was made with the forehand welding technique. The clean weld bead and lighter metal color immediately next to the bead demonstrate the cleaning and etching action of the reverse polarity welding current.

Goodheart-Willcox Publisher

Figure 10-13. Preheating the aluminum plate to a temperature between 200°F and 300°F (93°C and 149°C) helps prevent cold starts. The temperature-indicating crayon melts when the material reaches 250°F (121°C).

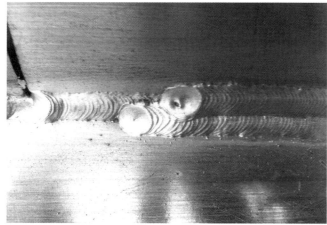

Goodheart-Willcox Publisher

Figure 10-16. The craters at the left end of this weld were formed when welding stopped. The craters exhibit cracks and porosity.

Goodheart-Willcox Publisher

Figure 10-17. The run-off tabs for this butt weld are approximately 3/4″ to 1″ wide and 3″ long.

Spot, Plug, and Slot Welding

For these types of welds, use the procedures outlined in this chapter and specified in the *Reference Section*. Welding techniques for these joints are the same as those used with carbon steels and stainless steel.

Burnback

Burnback of a welding electrode can be a serious and costly problem. Wire burnback occurs when the aluminum electrode melts and adheres to the contact tip. Although burnback can occur in GMAW in any welding mode and with any type of wire, the problem is greatest when aluminum is welded. This is because the metal's melting point is very low (1200°F or 647°C) and both the contact tip and electrode are nonferrous materials.

Burnback to the contact tip results in the electrode wire sticking in the tip while the drive rolls continue to push the wire. With aluminum, the result is birdnesting of the electrode, **Figure 10-18**. Burnback is costly to repair. In many cases, the tip is ruined beyond repair, and the electrode wire needs to be reinstalled through the gun. The following are hints for avoiding burnback when you are setting up or adjusting a procedure.

- Start with a close arc voltage setting and then adjust to the higher desired setting.
- Start with a higher wire feed setting and then adjust to the lower desired setting.
- Use very short arc times when starting. Carefully watch the arc until the correct values are established for the desired weld.
- Closely watch your wire supply. If the supply runs out during a weld, burnback will occur and possibly ruin the contact tip.
- Keep the wire supply system in top shape. Clean out the liners on a regular basis and replace liners as necessary. Check the drive rollers often and adjust them as necessary.

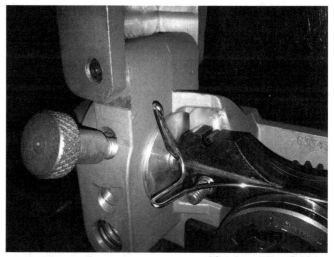

Goodheart-Willcox Publisher

Figure 10-18. Aluminum electrode wire has kinked and bent as the drive rolls continued to push the soft aluminum wire after the electrode melted to the contact tip. The wire must be removed and reinstalled.

Summary

- Aluminum alloys are nonferrous metals. They can be welded using the GMAW process in the short circuiting, spray transfer, and pulsed spray transfer modes.
- Characteristics of aluminum include high thermal conductivity, high electrical conductivity, high ductility at subzero temperatures, light weight, and high corrosion resistance. Aluminum is nonsparking, is nontoxic, and does not change color when heated.
- The aluminum family is made up of pure aluminum and various alloys. Aluminum alloys are separated into different series, based on the alloying elements added to the aluminum.
- Nonheat-treatable alloys are 1xxx, 3xxx, 5xxx, and some 4xxx series. The heat-treatable alloys include the 2xxx, 6xxx, 7xxx, and some 4xxx series. The heat-treated alloys are identified by temper designations.
- The two most commonly used electrode wires are 4043 and 5356. Several other electrode designations are available for specific alloys.
- Clean all aluminum oxide off the metal with a stainless steel brush prior to welding.
- Preheating aluminum reduces the amount of welding heat needed to achieve penetration and increases welding speed.
- Take precautions to prevent electrode wire burnback. This condition results in costly repair time and leads to other wire feed problems.

Review Questions

Answer the following questions using the information provided in this chapter.

1. List eight characteristics of aluminum.
2. Which series of alloying elements has the highest strength of the nonheat-treatable alloys?
3. Why should the square-groove butt weld design be avoided when aluminum is welded?
4. Common electrode wires used for welding aluminum are defined by the specification AWS _____.
5. Why must the oxidized metal produced by cutting or gouging be removed before aluminum is welded?
6. List two chemical means and two mechanical means of cleaning joints in preparation for welding.
7. How does adding helium to the shielding gas affect weld quality?
8. Preheating the base metal reduces the amount of welding heat needed to achieve penetration and increases welding _____.
9. How can welders avoid leaving a crater at the end of a weld pass?
10. What is meant by the term *burnback*?

An aluminum art piece that the author helped weld and build. The entire ball was GMAW aluminum welded to an aluminum plate and attached to a steel base.

Chapter
11
Gas Metal Arc Welding Procedures for Sheet Metal

Objectives

After studying this chapter, you will be able to:

- ❑ Explain the sheet metal gauge measurement system.
- ❑ Employ sheet metal welding procedures to reduce heat input and distortion.
- ❑ Spot weld sheet metal.

Technical Terms

burnback timer
intermittent (stitch) welds
sheet metal gauge
whisker

Sheet Metal

The GMAW process is ideally suited for welding sheet metals. Sheet metal is available in steel, stainless steel, and aluminum. All are widely used in HVAC, automotive and trailer, small parts, and food processing equipment fabrication processes. Sheet metals are also used in corrosive environment applications.

A gauge number identifies sheet metal according to the material thickness. See **Figure 11-1**. As the gauge number increases, the material thickness decreases. Steel, stainless steel, and aluminum gauge dimensions are each determined differently. A *sheet metal gauge* scale is used to check the thickness of the material, **Figure 11-2**. One side shows the decimal size, and the other shows the gauge number. See **Figure 11-3**.

Two major difficulties with using GMAW for sheet metal are distortion and burn-through. Procedure and equipment variables that can help eliminate these defects are discussed in the following sections.

Welding Procedures

The proper welding procedure is necessary to reduce weld heat input and the resulting distortion of the thin metal. The GMAW short circuiting transfer mode is best suited for welding thin-gauge sheet metal. However, a pulsed spray transfer mode is also suitable if travel speed is fast enough.

Sheet Metal Gauge Thickness				
Gauge	Mild Steel	Aluminum	Galvanized Steel	Stainless Steel
3	0.2391	0.2294		0.2500
4	0.2242	0.2043		0.2344
5	0.2092	0.1819		0.2187
6	0.1943	0.1620		0.2031
7	0.1793	0.1443		0.1875
8	0.1644	0.1285	0.1680	0.1650
9	0.1495	0.1144	0.1532	0.1562
10	0.1345	0.1019	0.1382	0.1406
11	0.1196	0.0907	0.1233	0.1250
12	0.1046	0.0808	0.1084	0.1094
13	0.0897	0.0720	0.0934	0.0937
14	0.0747	0.0641	0.0785	0.0781
15	0.0673	0.0571	0.0710	0.0703
16	0.0598	0.0508	0.0635	0.0625
17	0.0538	0.0453	0.0575	0.0562
18	0.0478	0.0403	0.0516	0.0500
19	0.0418	0.0359	0.0456	0.0437
20	0.0359	0.0320	0.0396	0.0375
21	0.0329	0.0285	0.0366	0.0344
22	0.0299	0.0253	0.0336	0.0312
23	0.0269	0.0226	0.0306	0.0281
24	0.0239	0.0201	0.0276	0.0250
25	0.0209	0.0179	0.0247	0.0219
26	0.0179	0.0159	0.0217	0.0187
27	0.0164	0.0142	0.0202	0.0172
28	0.0149	0.0126	0.0187	0.0156
29	0.0135	0.0113	0.0172	0.0141
30	0.0120	0.0100	0.0157	0.0125
31	0.0105	0.0089	0.0142	0.0109
32	0.0097	0.0080	0.0134	0.0102
33	0.0090	0.0071		0.0094
34	0.0082	0.0063		0.0086
35	0.0075	0.0056		0.0078
36	0.0067			0.0070

Goodheart-Willcox Publisher

Figure 11-1. Standard gauge sizes (in decimal inches) of various types of sheet metal. As the gauge number increases, the material thickness decreases.

A

B

Goodheart-Willcox Publisher

Figure 11-2. Sheet metal gauge. A—Decimal equivalent side of a steel sheet metal gauge. B—Standard gauge number side of a steel sheet metal gauge.

Goodheart-Willcox Publisher

Figure 11-3. This steel sheet metal gauge is being used to verify 16-gauge material.

Decrease electrode wire size and wire feed speed to reduce heat input if these changes still allow the weld to be completed with a single bead. Shielding gas should include a lower percentage of carbon dioxide when mild steel sheet metal is welded.

Sheet metal distortion occurs when the metal heats up and causes expansion of the material. The thin sheet metal bends, buckles, and eventually burns through when the welding heat is concentrated in one place. See **Figure 11-4** and **Figure 11-5**. Follow these guidelines to reduce distortion:

- Use clamps to securely hold the metal, **Figure 11-6**.
- Use several small tack welds to hold the sheet in place.

- Use short weld beads of one to two inches, depending on sheet gauge. These short beads are commonly referred to as ***intermittent (stitch) welds***. See **Figure 11-7**. Stagger the intermittent welds from one area to another, allowing the metal to cool before placing an adjacent weld. See **Figure 11-8**. Repeat the process to fill gaps between the intermittent welds, creating a continuous weld bead.
- Use a backing chill bar or plate to absorb welding heat away from the sheet metal.
- Cool the surrounding metal (not the weld) with a cool wet cloth after each weld.
- Bevel butt welds to flatten bead profile and increase penetration.

Goodheart-Willcox Publisher

Figure 11-4. A welded pan with distortion, buckling, and poor fitup caused by excessive heat and poor welding procedures.

Goodheart-Willcox Publisher

Figure 11-6. Sheet metal properly clamped prior to welding.

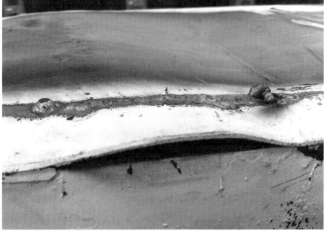

Goodheart-Willcox Publisher

Figure 11-5. The bottom side of a welded pan shows distortion and burn-through caused by excessive heat input and improper welding procedures.

Goodheart-Willcox Publisher

Figure 11-7. Using the short circuiting mode of metal deposition, a forehand direction, reduced voltage and amperage along with short, intermittent weld beads will reduce heat input and distortion.

Goodheart-Willcox Publisher

Figure 11-8. The bottom side of the tray welded with proper welding procedures. The weld will be completed after this section cools.

Forehand welding and maintaining a short contact tip to work distance reduces weld burn-through. Alternating the gun trigger on and off to create short welds also reduces heat input and burn-through.

Spot Welding Sheet Metal

Some applications do not require a continuous weld bead. In such cases, spot welding can be the primary method of joining components. Spot welding can also be used to hold component parts assembled together for the final welding operation. A variety of metals and joint types can be spot-welded. Spot welding is commonly used to weld lap joints. See **Figure 11-9**.

The spot welding process requires a welding machine with special timers installed to control the welding operation timing sequence. A typical welding power source for thin-gauge materials has a timer assembly installed. A timer assembly typically includes the following:

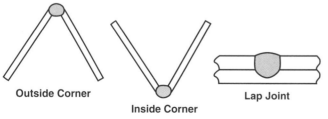

Goodheart-Willcox Publisher

Figure 11-9. Joints that are spot welded include corner joints and lap joints.

- **Welding timer.** This timer controls the length of time the welding current is on.
- **Shielding gas postflow timer.** This timer controls the length of time the gas flows after the arc is stopped. Usually the gas should flow for several seconds after the weld is stopped to prevent contamination of the molten metal.
- **Anti-stick timer.** This timer is commonly called a *burnback timer*. The timer is placed into the welding circuit to allow the electrical current to flow to the electrode wire after the electrode wire stops feeding. The flow of current continues to "burn back" the welding wire to prevent the wire end from freezing into the molten pool.

The welding gun requires special nozzles with slots for the escape of welding heat, gases, and air during the welding operation. Each gun manufacturer designs special nozzles for these types of operations. The nozzles cannot be used on guns made by other manufacturers.

Preweld Cleaning and Fitup

Lap joints on sheet metal often require special cleaning procedures if the weld will be used in applications where strength is required. The area where the two metals contact must be clean. Because the heat for fusion must come through the upper material, any scale or rust prevents fusion between the two plates. Materials should be cleaned with a sander, sandblasted, or abrasively cleaned to bright metal. Any oil on the surface must be removed prior to welding.

Another problem that may be encountered is gapping of the two plates. If the plates are gapped, molten metal will cool before fusion is complete with the bottom plate. Insufficient fusion lowers the strength of the weld. The use of multiple clamps and a heavy blow with a hammer on the weld after each tack will reduce gapping.

When spot welds are made on lap joints of different thicknesses, the thinnest material must always be on the top of the joint. If the design of the assembly cannot be changed, a plug weld can be made in place of the spot weld.

Where spot welds are used on inner or outer corner joints, the fitup gap of the mating edges must be minimized, or burn-through will occur. If the welding wire passes through the gap, it will burn off and become a *whisker* (a piece of electrode wire that extends through the completed weld joint) on the root side of the weld.

Summary

- Short circuit gas metal arc welding is ideal for sheet metal welding applications.
- Sheet metal thickness is identified by the gauge size. The higher the gauge number, the thinner the material.
- The major difficulties in welding sheet metal are excessive heat input and the resulting material distortion and burn-through.
- Proper machine setup and welding procedures should be followed to reduce welding heat and distortion in order to produce quality welds.
- Spot welding is commonly used to weld lap joints. Additional controls are often installed in the wire feeder for spot welding.

Review Questions

Answer the following questions using the information provided in this chapter.

1. Which of the following is the thinnest sheet metal—10-gauge, 16-gauge, or 24-gauge?

2. The two major problems encountered with gas metal arc welding of sheet metal are burn-through and _____.

3. Thin-gauge sheet metal can be welded in the pulsed spray transfer mode if the travel speed is _____ enough.

4. The use of a smaller _____ and reduced wire feed speed will decrease welding amperage and weld heat input.

5. What is the cause of sheet metal distortion?

6. Short weld beads, commonly called _____, can be used to reduce distortion.

7. List two ways to help absorb the welding heat input in sheet metal.

8. Although they can also be used on outside and inside corner joints, spot welds are most commonly used on _____ joints.

9. What is the purpose of an anti-stick (burnback) timer?

10. When a spot weld is made on sheet metal of different thicknesses, the _____ metal should always be on the top.

Goodheart-Willcox Publisher

This student has just successfully completed a test weld.

Flux Cored Arc Welding Procedures and Techniques

Objectives

After studying this chapter, you will be able to:

❏ Determine FCAW procedure requirements prior to welding.

❏ Use appropriate welding techniques for specific conditions.

❏ Make test welds and record the data on a weld schedule.

❏ Troubleshoot common FCAW problems that occur during initial setup and production.

Technical Terms

backhand welding
drag angles
forehand welding
stringer bead pattern
undercut
visible stickout
weave bead pattern

Flux Cored Arc Welding Operation

Flux cored arc welding (FCAW) can be either a semiautomatic or fully automatic operation. In semiautomatic welding, technique is critical to the quality of the weld. The welder must be able to set up the machine and operate the gun properly to produce a weld that meets strict fabrication requirements. In a fully automatic operation, the machine program controls the welding parameters; however, operator skill remains important to ensure that weld quality is maintained.

Welding Procedure Requirements

Before any type of welding is performed, a welding procedure should be developed and tested to establish the filler materials for the specific base materials, prove the joint design, and obtain the actual welding parameters. Once the welding procedure is developed, all welders should follow the procedure precisely. A separate welding procedure must be developed for any variations in essential variables, such as the type of material being welded or the joint design.

Base Materials and Joint Types

Base materials are indicated by specification, grade, type, and thickness. Joint design provides details on the type of joint, including bevel angles, root spacing, type of backing material, and specifications for completing the required weld.

Electrodes

Different FCAW electrodes are made for out-of-position welds; for high-deposition rates for groove welds or build-up–type welds; and for welding stainless steels, cast irons, or various grades of carbon steel. Two different types of flux cored electrode wires—dual-shielded and self-shielded—are used in FCAW. These electrodes had been classified under AWS A5.20/A5.20M and AWS A5.29/A5.29M until 2012. At that time, AWS published AWS A5.36/A5.36M *Specifications for Carbon and Low-Alloy Steel Flux Cored Electrodes for Flux Cored Arc Welding and Metal Cored Electrodes for Gas Metal Arc Welding.* This open classification system was created to address changes in cored electrodes.

FCAW-G is a dual-shielded electrode that provides shielding using the flux supplied within the cored wire but requires use of additional shielding gas. Some dual-shielded electrodes are identified as E71T-1C. The *C* indicates that the electrode should be used with 100% CO_2 shielding gas. Other dual-shielded electrodes are identified as E71T-1M. The *M* indicates that the electrode should be used with a mixed shielding gas consisting of 75%–80% argon and 20%–25% CO_2. See **Figure 12-1**.

FCAW-S is a self-shielded electrode wire that is used without any additional shielding gas. When the letter *C* or *M* is not included in the AWS designator, the electrode is self-shielded. The chemical reaction of the flux with the atmosphere is required to allow the flux to work properly as it hardens into the slag deposit. An example of this type of electrode is E71T-3. See **Figure 12-2**.

Electrode diameters range from 0.030″ to approximately 0.150″ and are available in fractional and decimal sizes. In general, the smaller electrodes are used for out-of-position welding, and the larger electrodes are used for flat and horizontal welding. For

Used with permission of the Lincoln Electric Company, Cleveland, Ohio, USA

Figure 12-2. Various types and sizes of FCAW-S electrodes.

the lowest arc time and cost, use the largest possible electrode size.

With a constant voltage (CV) power source and a preset arc voltage, electrode wire feed speed determines the amount of welding current. With a constant current (CC) power source and a voltage-sensing wire feeder, the welding current and electrode wire speed fed into the molten pool determine the amount of arc voltage. Electrode feed is set using the manufacturer's setup graph or formula. **Figure 12-3** shows a voltage-sensing wire feeder. The voltage-sensing lead will be attached to the work to supply electrical feedback to the wire feeder, allowing it to maintain the voltage required during the welding process.

Feed rate control

Gun and cable connection

Voltage-sensing lead

Welding machine connection

Used with permission of The Lincoln Electric Company, Cleveland, Ohio, USA

Figure 12-3. A voltage-sensing wire feeder that will be attached to a diesel engine powered welding machine. This type of wire feeder is commonly called a *suitcase feeder.*

Figure 12-1. Various types and sizes of FCAW-G electrodes.

Current Type and Polarity

Polarity is set on the power source (if a switch is available), or the electrode leads are connected to the appropriate output terminals. For DCEN polarity, also referred to as DCSP (straight polarity), the power cable is connected to the negative terminal, and the work cable is connected to the positive terminal. Most FCAW-S electrodes, with a few exceptions, use DCEN polarity. For DCEP polarity, also referred to as DCRP (reverse polarity), the power cable is connected to the positive terminal, and the work cable is connected to the negative terminal. All FCAW-G uses DCEP.

If a CC wire feeder is being used, the cables must be properly connected to the power source, and the polarity switch must be in the correct position for the unit to operate. These units are designed to shut down if connections are not properly made. The work cables connect the power source and the workpiece.

Shielding Gas

The electrode manufacturer specifies the shielding gas type. Shielding gas is important to the deposition of the molten metal and final metallurgical properties of the deposit. The gas nozzle and gas flow rate must be properly selected. Flow rates range from 35 cfh to 50 cfh for welding in still air. Avoid welding in windy or drafty areas where gas coverage is poor, or switch to a self-shielded electrode.

Electrode Extension

When measuring electrode extension, it is important to measure the distance from the contact tip to the end of the electrode and not just the amount of electrode extending from the end of the gas nozzle. Each electrode manufacturer specifies electrode extension distance, which the welding operator must maintain. See **Figure 12-4**. Electrode extension for FCAW-G is slightly longer than that used for GMAW. This distance allows the shielding gas to cover the flux as it is formed. If this distance is not maintained, the gas becomes trapped beneath the weld metal and the flux. This trapped gas results in defects, **Figure 12-5**.

When long electrode extensions (and long nozzles) are used for high-deposition welding, the contact tip is not visible to the operator. *Visible stickout* refers the distance between the end of the gas nozzle and the end of the electrode that can be viewed. See **Figure 12-6**.

Increasing the extension lowers the current at the weld pool and lowers the voltage across the arc. Lowering arc voltage increases bead convexity (height of crown), decreases porosity, and reduces penetration. Decreasing the extension increases weld current at the molten pool and increases penetration.

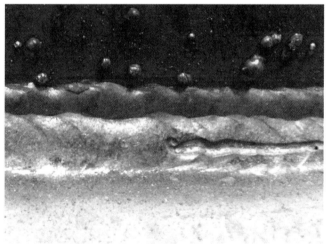

Goodheart-Willcox Publisher

Figure 12-5. When the electrode extension distance is too close or when an excessive weld drag angle is used, the shielding gas is trapped beneath the slag covering and forms gas pockets.

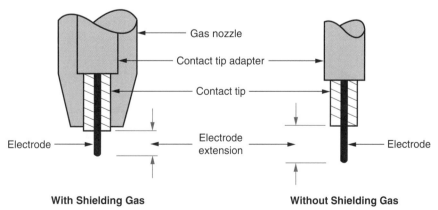

With Shielding Gas   **Without Shielding Gas**

Goodheart-Willcox Publisher

Figure 12-4. The welder controls electrode extension during the welding operation. To maintain an even flow of current to the electrode, hold the extension distance as closely as possible to the manufacturer's recommendation.

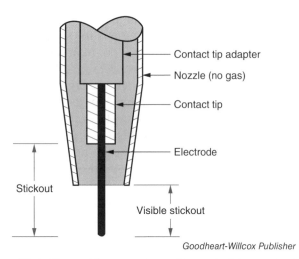

Goodheart-Willcox Publisher

Figure 12-6. The welder cannot see the end of the contact tip. Electrode visible stickout is measured from the end of the special nozzle to the end of the electrode.

Welding Techniques

In an automatic welding operation, travel speed does not vary like it does in a semiautomatic operation. In the semiautomatic mode, the welder can adjust speed and other variables to produce a satisfactory weld. Practice in making all types of joints, especially those requiring multi-pass welds, is essential to developing proper technique. Every pass affects the weld. Traveling too slowly produces wide welds, while traveling too quickly reduces width.

In *backhand welding* (pull welding), the welding gun is pointing back at the weld and the weld progresses in the opposite direction. See **Figure 12-7**. In *forehand welding* (push welding), the welding gun points forward in the direction of travel. The forehand gun angle is not often used in the flat position because slag tends to flow forward and become trapped in the weld. For welding in the vertical position, both the forehand and backhand techniques can be used, **Figure 12-8**. Overhead position welding is generally done with the backhand technique.

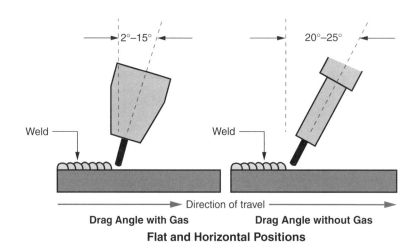

Goodheart-Willcox Publisher

Figure 12-7. Penetration into the base metal is deeper with a slight gun angle. Weld crown height increases as the gun angle is increased.

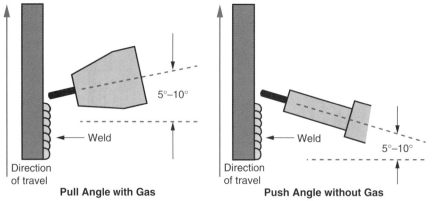

Goodheart-Willcox Publisher

Figure 12-8. Tilting the gun into a pull angle keeps the shielding gas on the molten pool. When welding is performed with a self-shielding electrode in the vertical position, the push technique can be used.

Gun Angles

Gun position is described by a combination of the gun's longitudinal (along the weld) angle and its transverse (across the weld) angle. The longitudinal angle is often referred to as the travel angle and the transverse angle is referred to as the work angle. See **Figure 12-9**. Travel angles for backhand welding on linear welds are called *drag angles* and have a considerable effect on penetration, bead form, and final weld bead appearance. As drag angle increases, the weld penetration decreases and the bead crown height increases. If the drag angle is excessive, the bead will appear to sit on top of the base metal. Excessive drag angle also increases the chance of trapping shielding gas under the flux when FCAW-G is used. Adjust gun angle and travel speed to maintain good weld pool control and weld bead shape.

FCAW electrodes contain different amounts of flux. Therefore, gun angles vary from slight (for small amounts of flux deposition) to considerable (for electrodes with substantial flux content). If the drag angle is insufficient, the flux moves around to the front of the pool and becomes entrapped in the weld.

The work angles needed for proper bead placement on flat and horizontal groove welds are shown in **Figure 12-10** and **Figure 12-11**. These angles and bead placement positions may also be used for overhead welding. In the horizontal and overhead positions, weld metal tends to flow downward, causing poor bead formation. Such welds should be made with stringer beads for better control of the weld metal and final bead surface contour.

Bead Sequence

When large fillet welds are required, make multiple-pass welds and start the bead sequence as shown in **Figure 12-12**. Work upward, adding passes

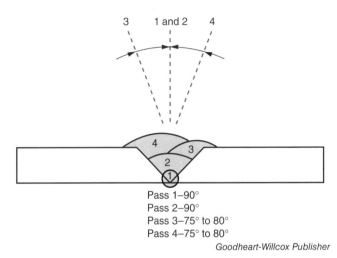

Pass 1–90°
Pass 2–90°
Pass 3–75° to 80°
Pass 4–75° to 80°

Goodheart-Willcox Publisher

Figure 12-10. This sequence of bead deposition is used for heavy joints. Always deposit weld metal to permit sufficient area for the next pass.

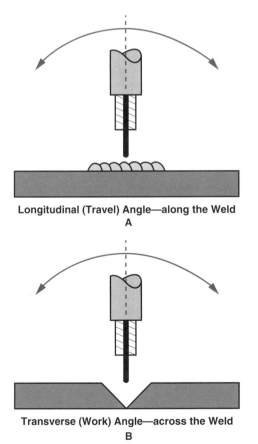

Longitudinal (Travel) Angle—along the Weld
A

Transverse (Work) Angle—across the Weld
B

Goodheart-Willcox Publisher

Figure 12-9. A—The travel angle is the gun's angle along the weld's axis. B—The work angle is the gun's angle across the weld axis.

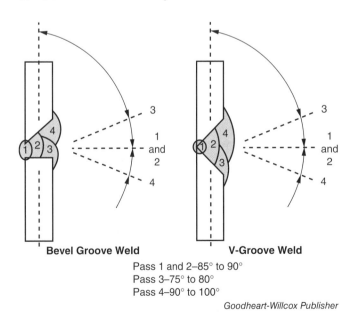

Bevel Groove Weld   **V-Groove Weld**

Pass 1 and 2–85° to 90°
Pass 3–75° to 80°
Pass 4–90° to 100°

Goodheart-Willcox Publisher

Figure 12-11. Horizontal welds tend to fall when too much metal is deposited on the weld pass. Move faster and make stringer beads to control bead formation. Fill the groove from the bottom up.

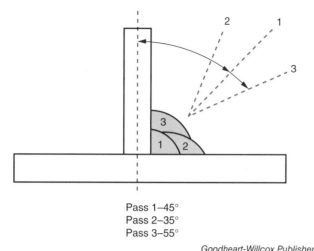

Pass 1–45°
Pass 2–35°
Pass 3–55°

Goodheart-Willcox Publisher

Figure 12-12. Large horizontal fillets require more than one pass. Do not try to make the weld in one or two passes.

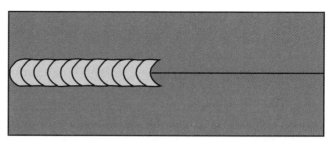

Goodheart-Willcox Publisher

Figure 12-13. A stringer bead pattern is made with little or no side-to-side motion.

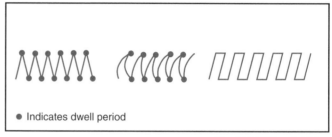

● Indicates dwell period

Goodheart-Willcox Publisher

Figure 12-14. Types of weave bead patterns.

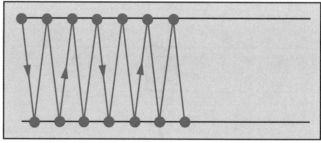

Goodheart-Willcox Publisher

Figure 12-15. Oscillated wash beads are made to cover previous welds and require a dwell at the end of the weave to prevent undercut.

as required. When welding uphill with a gas-shielded electrode, use a slight drag angle so the gas covers the molten pool. When using a self-shielded electrode in the vertical position, use a slight push angle for better slag control and visibility of the molten pool.

Both stringer and weave beads are used for welding uphill. Successful vertical welds depend on several factors, including the type of electrode, operator ability, desired weld quality, and final crown shape. Welding downhill is not recommended for FCAW-G and FCAW-S due to the possibility of slag entrapment in the weld. If downhill welding is necessary, the travel angle must be a drag-type leading angle. Insufficient travel angle permits the slag to run into the molten pool, resulting in slag pockets or incomplete penetration.

Bead Patterns

With a *stringer bead pattern*, travel is nearly straight along the joint. In this pattern, there is very little side-to-side motion. A small zigzag motion or small circular motion is used. See **Figure 12-13**. The weave is made only enough to prevent slag entrapment in the weld and ensure that the molten metal washes into the previous weld or base material.

A *weave bead pattern* (also referred to as wash beads or oscillated beads) flows metal over a wider area, **Figure 12-14**. Because travel speed is slower, weave beads create more heat than stringer beads, resulting in greater distortion. A depression (*undercut*) forms at the edges of the bead if a dwell (wait period) is not allowed at the end of each side-to-side motion. See **Figure 12-15**. The amount of forward movement at each dwell can be determined by closely watching

the metal fill at the edge of the weld and the height of the bead.

When an uphill vertical fillet weld is made on heavy plate, a triangular weave bead pattern is often used to weld the first pass of the multipass weld. Because this technique generates considerable heat, it cannot be used on thin materials. Considerable practice is required to be able to deposit the weld metal in consistent layers with adequate penetration. See **Figure 12-16**.

Test Welds

The welding of component parts involves a combination of parameters and variables, including

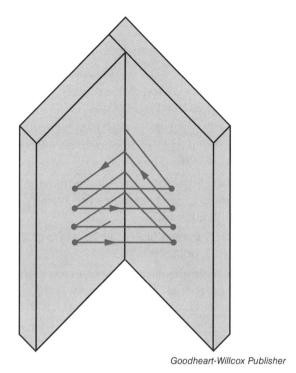

Goodheart-Willcox Publisher

Figure 12-16. Triangular weave patterns require practice to correctly time upward movements and deposit each layer in the proper place.

materials, wire sizes, types of gases, machine settings, and welding techniques. The proposed welding procedure must be tested before the actual weld is made. To determine the actual values, select an initial setting for the type and thickness of material involved. Settings are found in the operating parameter charts in the *Reference Section*.

Prepare the joint (if required), clean the material, and make a series of test welds. During the test period, adjust the machine settings and welding techniques until the desired weld is produced. When the test weld meets all the visual and mechanical requirements of the specification, record the parameters and variables on a weld schedule. Refer to **Figure 7-14**. Recording the data for a particular weld enables the weld to be duplicated at a future date.

A weld is like a signature—unique to every individual. Each welder has his or her own technique, meaning no two welds are exactly alike. You sign your work, so do your best on any job. Because machine welds can be duplicated, the tolerance for variables can be smaller or tighter.

During initial setup and production, problems can occur that affect the FCAW operation. Many of these problems can be corrected with simple changes, as outlined in **Figure 12-17**.

FCAW Troubleshooting Chart		
Problem	**Possible causes**	**How to correct**
Electrode stubs against work.	Voltage is too low. Electrode feed is too high. Loose ground connection.	Raise (increase) voltage. Decrease electrode feed. Change or tighten ground.
Electrode burns back into contact tip.	Voltage is too high. Electrode feed speed is too low. Stickout is too short.	Decrease voltage. Increase electrode feed speed. Increase stickout.
Electrode feed is erratic.	Kinked electrode feed conduit. Electrode feed drive roller slipping. Kinked filler electrode. Brake adjustment incorrect. Contact tip partially clogged.	Straighten. Adjust idler roll. Check filler electrode for kinks. Check brake tension. Clean ream hole.
Arc blows metal from intended path.	Poor ground location. Insufficient ground clamp area.	Change ground location. Add another ground.
Weld pool extremely fluid.	Current too high.	Reduce electrode speed or amperage.
Weld pool very sluggish.	Current too low. Stickout too long.	Increase electrode speed or amperage. Decrease stickout.
Arc outages.	Contact tip hole oversize. Poor ground connection.	Change contact tip. Change workpiece connection.

Goodheart-Willcox Publisher

Figure 12-17. A troubleshooting guide for operational problems that may occur during FCAW.

Summary

- FCAW can be either a semiautomatic or fully automatic operation.
- FCAW-G is a dual-shielded electrode that provides shielding using the flux supplied within the cored wire but requires use of additional shielding gas. The shielding gas is either 100% CO_2 or a mixture of 75%–80% argon and 20%–25% CO_2 as a shielding gas.
- FCAW-S is a self-shielded electrode wire that is used without a shielding gas. The chemical reaction of the flux with the atmosphere is required for the flux to work properly.
- Different FCAW electrodes are made for out-of-position welds; high-deposition rates for groove welds or build-up–type welds; and for welding stainless steels, cast irons, or various grades of carbon steel.
- FCAW electrodes contain different amounts of flux. Travel angles vary from slight (for small amounts of flux deposition) to considerable (for electrodes with substantial flux content).
- Electrode extension for FCAW-G is slightly longer than that used for GMAW in order to allow the shielding gas to cover the flux as it is formed.
- Welding downhill is not recommended for FCAW-G and FCAW-S due to the possibility of slag entrapment in the weld.
- With a stringer bead pattern, travel is nearly straight along the joint. A weave bead pattern flows metal over a wider area.
- The welding of component parts results from a combination of parameters and variables, including materials, wire sizes, types of gases, machine settings, and welding techniques. The proposed welding procedure must be tested before the actual weld is made.

Review Questions

Answer the following questions using the information provided in this chapter.

1. What are the two different types of cored electrode wires used in FCAW?
2. *True or False?* A dual-shielded electrode that is identified as E71T-1C should be used with a mixed shielding gas.
3. With a CV power source and a preset arc voltage, electrode _____ determines the amount of welding current.
4. With few exceptions, most FCAW-S electrodes use _____ polarity.
5. Decreasing the electrode extension increases weld current at the molten pool and _____ penetration.
6. Why is the forehand welding technique *not* often used in the flat position?
7. *True or False?* A slight travel angle is used with electrodes that contain a small amount of flux.
8. In a weave bead pattern, a(n) _____ forms at the edge of the weld if a dwell is not allowed at the end of each side-to-side motion.
9. A triangular weave bead pattern *cannot* be used on thin materials because it generates considerable _____.
10. According to the troubleshooting guide in this chapter, what are the causes of a very sluggish weld pool?

Flux Cored Arc Welding Procedures for Carbon Steels

Objectives

After studying this chapter, you will be able to:

❏ Identify carbon steels that are welded with the FCAW process.

❏ Select the appropriate electrode wire based on usability and performance capabilities.

❏ Prepare a carbon steel joint for welding.

❏ Determine the welding procedures required to produce a quality weld.

❏ Make a test weld on carbon steel using the FCAW process.

Technical Terms

carbon steels
interpass temperature
low-carbon steel
postheating
preheating

Carbon Steels Welded with FCAW

As discussed in Chapter 8, *carbon steels* are a group of steels that contain carbon, manganese, and silicon, along with small amounts of other elements. Alloy steel is steel that has a variety of other elements of up to 50% by weight to improve its mechanical or physical properties. Alloy steels are broken down into two groups—low-alloy steels and high-alloy steels.

Steels that can be welded with the FCAW process include the following:

- Mild and structural steel, identified as ASTM A36, A515, and A516 grades.
- High-strength, low-alloy steels. These alloy steels provide better mechanical properties or greater resistance to corrosion than carbon steel. They are made to specific mechanical properties rather than specific chemical compositions.
- High-strength, quenched and tempered alloy steels. A higher strength, abrasion-resistant and tougher steel that has been tempered to achieve weldability.
- Carbon-molybdenum steels. This grade of steel is used almost exclusively in refinery and petrochemical applications due to the special alloys that are chemically resistant.
- Chromium-molybdenum (chrome-moly) steels, such as 1.25% Cr–0.5% Mo and 2.25% Cr–1.0% Mo. This is a range of low alloy steels used to produce tubing with high tensile strength and malleability. It is easily welded and considerably stronger than mild steel tubing.

- Nickel steels. An alloyed steel with nickel as the main alloy provides a high strength, ductile, and tough steel for use in acidic environments.
- Manganese-molybdenum steels. Manganese is used as an alloy element to improve hot ductility.

Electrodes

Electrode wires for welding carbon steels are identified in AWS specification A5.36/A5.36M *Specification for Carbon and Low-Alloy Steel Flux Cored Electrodes for Flux Cored Arc Welding and Metal Cored Electrodes for Gas Metal Arc Welding*. Each electrode is identified by a series of letters and numbers that corresponds to its strength, chemical composition, and usability. Refer to Chapter 5 to review identification of electrodes for carbon and low-alloy steels.

Electrode Characteristics

Manufacturers design different types of electrodes for use with different metal transfer modes. In the spray transfer mode, weld metal is deposited as small droplets of metal surrounded by slag and smoke. In contrast, the globular transfer mode consists of larger globules of metal that detach from the wire in an indefinite pattern.

In the electrode classification system, usability and performance capabilities are indicated after the letter *T* (for tubular electrode). An E71T-1 wire, for example, specifies electrode classification T-1. See **Figure 13-1**.

FCAW Electrode Classifications	
Usability Designator	**Characteristics**
T-1	DCEP. Dual-shield CO_2 or Ar/CO_2 shielding gas. Single or multiple-pass welds. Short circuiting transfer or spray transfer with low spatter loss and moderate slag.
T-2	DCEP. Dual-shield CO_2 shielding gas. Single-pass welding in flat and horizontal fillet weld positions. Tolerates some mill scale and oxidation on base metal. Arc is similar to T-1.
T-3	DCEP. Self-shielding—No shielding gas. Spray transfer. Single-pass welds; not for use on multiple-pass welds.
T-4	DCEP. Self-shielding. Globular transfer. High deposition rate on poor fitup, shallow penetration on flat and horizontal positions. Single- or multiple-pass welds.
T-5	DCEP. Dual-shield CO_2 or Ar/CO_2 shielding gas. Globular transfer. Low-hydrogen electrode similar to SMAW E7018 electrode. Excellent crack-resistance and impact properties.
T-6	DCEP. Self-shielding. Spray transfer. Weld has good impact properties, deep penetration, and excellent slag removal. Single- or multiple-pass welds in flat or horizontal positions.

Figure 13-1. FCAW electrode classifications.

(Continued)

FCAW Electrode Classifications *(Continued)*	
Usability Designator	**Characteristics**
T-7	DCEP. Self-shielding. Small diameter electrodes for all position welds, large diameter electrodes for flat and horizontal position welds. Slag system designed to desulfurize the weld metal, producing a crack-resistant deposit.
T-8	DCEN. Self-shielding. All positions, single- or multiple-pass welds. Weld is crack-resistant with good impact properties.
T-9	DCEP. Dual-shield CO_2 or Ar/CO_2 shielding gas. Single- or multiple-pass welds. Weld is crack-resistant with good impact properties.
T-10	DCEN. Self-shielding. Single-pass welds on material of any thickness in flat or horizontal position.
T-11	DCEN. Self-shielding. Spray transfer. Single- or multiple-pass welds in all positions. Material thickness limitations. Preheat, interpass, and postheat temperature requirements for multiple-pass welds.
T-12	DCEP. Dual-shield CO_2 or 75%Ar/25% CO_2 shielding gas. Single- or multiple-pass welds. Weld is crack-resistant with good impact properties.
T-13	DCEN. Self-shielding. Single-pass welds. Impact values not specified.
T-14	DCEN. Self-shielding. Single-pass welds. Impact values not specified.
T-G	See manufacturer's electrode data for information.
T-GS	See manufacturer's electrode data for information.

Goodheart-Willcox Publisher

Figure 13-1. *(Continued)*

Chemical and Mechanical Values of Flux Cored Arc Welds

Filler material specifications require that welding tests be completed by the manufacturer before acceptance of each lot of FCAW electrodes. The tests are performed with qualified welding procedures on a specific type of material. The deposited metal must meet certain chemical, mechanical, and other requirements to meet the specifications. See **Figure 13-2** and **Figure 13-3**. Actual test results of each lot or heat are available to purchasers of electrodes.

In comparison to mild steel electrode wire, electrodes manufactured for welding different types of alloy steels have added alloys in the wire content. Electrodes designated by major alloy content are as follows:
- Carbon-molybdenum steel electrodes have additional molybdenum content.
- Chromium-molybdenum steel electrodes have additional chromium and molybdenum content.

Typical Electrode Chemical Composition										
	Carbon (C)	Phosphorus (P)	Sulfur (S)	Vanadium (Va)	Silicon (Si)	Nickel (Ni)	Chromium (Cr)	Manganese (Mg)	Molybdenum (Mo)	Aluminum (Al)
T-1 T-4 T-5 T-7 T-8 T-11 T-G	X	0.04	0.03	0.08	0.90	0.50	0.20	0.30	1.75	1.8
X = amount to be determined										

Goodheart-Willcox Publisher

Figure 13-2. Typical chemical analysis for carbon steel deposited weld metal.

Electrode Mechanical Properties			
Electrode Classification	Tensile Strength (ksi)	Yield Strength (ksi)	Percent Elongation 2″
E6XT-1	62	50	22
E6XT-4	62	50	22
E6XT-5	62	50	22
E6XT-6	62	50	22
E6XT-7	62	50	22
E6XT-8	62	50	22
E6XT-11	62	50	22
E6XT-G	62	50	22
E7XT-1	72	60	22
E7XT-2	72	NR	NR
E7XT-3	72	NR	NR
E7XT-4	72	60	22
E7XT-5	72	60	22
E7XT-6	72	60	22
E7XT-7	72	60	22
E7XT-8	72	60	22
E7XT-10	72	NR	NR
E7XT-11	72	60	22
NR = No Requirements			

Goodheart-Willcox Publisher

Figure 13-3. Typical values for electrode mechanical properties for carbon steel deposited weld metal.

- Nickel steel electrodes have a higher manganese content and additional nickel, chromium, molybdenum, vanadium, and aluminum.
- Manganese-molybdenum steel electrodes have additional molybdenum content.

Proper electrode selection requires matching the type of electrode to the specific base material to obtain

Electrode Mechanical Properties			
Electrode Classification	Tensile Strength (ksi)	Yield Strength (ksi)	Percent Elongation 2″
E6XTX-X	6-80	50	22
E7XTX-X	70-90	58	20
E8XTX-X	80-100	68	19
E9XTX-X	90-110	78	17
E10XT-X	100-120	88	16
E11XT-X	110-130	98	15
E12XT-X	120-140	108	14

Goodheart-Willcox Publisher

Figure 13-4. Typical values for electrode mechanical properties for low-alloy steel deposited weld metal.

the desired chemical and mechanical properties. See **Figure 13-4**. When selecting electrodes for welding on alloy steel, evaluate all the welding and metallurgical aspects and make an appropriate welding test.

Material Preparation

FCAW can often be performed on slightly rusted material without adversely affecting the completed weld. However, in many cases, specifications prohibit the use of rusted material because defects may form. Plate material should be clean and free of excessive rust, scale, grease, and oil. Weld joint preparation by the thermal-cutting process leaves an oxide film on the surface, which can cause defects. Scale and rust can be removed with sandblasting. Wire brushing or grinding with a power grinder are other methods used, **Figure 13-5**. Wear protective eye and face gear when operating scale or rust removal equipment.

Goodheart-Willcox Publisher

Figure 13-5. Clean weld base metal by grinding, sanding, or wire brushing to remove oxidation.

Welding Procedures

A joint welding procedure is a combination of joint design and proven welding parameters. Welding electrode manufacturers and others have established welding procedures for various joint types. Use established procedures and test results for similar joints to help you determine an initial procedure for welding a particular joint. As you produce the actual joint, consider how you could change various parameters to reduce cost or improve the quality of the weld. Major considerations in the establishment of a welding procedure are described in the following sections.

Preheat, Interpass, and Postheat

Preheating is heating the entire weld area to a specific temperature before welding begins. The temperature should be consistent throughout the thickness of the joint. *Interpass temperature* is the minimum/maximum temperature of the weld metal before the next pass in a multiple-pass weld is made. Always maintain interpass temperature until the postheating operation is started. *Postheating* is the final duration and temperature of heating after welding is completed and before the weld is allowed to cool to room temperature.

With *low-carbon steels* (carbon steels that contain up to 0.14% carbon) and mild steels, the amount of preheat, interpass temperature, and postheat is not critical unless the material is more than 1″ thick, has a severe joint restraint, or is very cold. Low-carbon steels do not have sufficient carbon and alloying elements

to cause hardening of the weld or the weld zone, so cracking is not a problem. As the carbon and alloy content increases, so does the possibility of cracking, and heating temperatures must be considered before, during, and after welding. See **Figure 13-6**. When welding with alloy electrodes or the higher-carbon or alloy steels, consult the manufacturer of the material or the electrode for preheat and interpass temperatures. See the table of preheat, interpass, and postheat temperatures in the *Reference Section*.

Several methods can be used to measure temperature, including temperature indicating guns, crayons, and paints. The easiest method is to use an indicating gun or crayon that melts at the desired temperature.

Joint Preparation

Joint designs for FCAW follow the same basic design as those for GMAW. Although both gas-shielded and self-shielded operations can be used, gas-shielded electrodes obtain better penetration because they result in higher current densities. See **Figure 13-7**. In joints to be welded with FCAW-G, groove angles can be smaller, or the groove joint may have a narrower root opening.

When designing the joint, consider the movement of the molten metal needed to prevent undercut. Also consider the visual access of the joint by the welder. The final joint design should be verified by a test to establish the welding parameters to be used for production welding.

A fillet weld made with gas shielding can be made smaller but still have the same strength as a self-shielded weld. This can result in considerable cost savings. See **Figure 13-8**.

Plate Thickness in. (mm)	Up to ¾ (19)	¾–1½ (19–38)	1½–2½ (38–64)	Over 2½ (64)
Recommended minimum preheat temperature, °F (°C)	70 (21)	150 (66)	150 (66)	225 (107)
Recommended interpass temperature, °F (°C)	70 (21)	150 (66)	225 (107)	300 (149)

Goodheart-Willcox Publisher

Figure 13-6. Recommended initial temperatures for preheat and interpass. Higher temperatures may be required depending on job conditions, codes, or the presence of cracks.

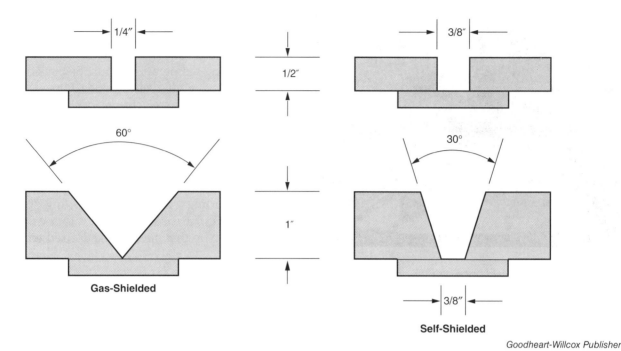

Goodheart-Willcox Publisher

Figure 13-7. Types of joint designs used with gas-shielded and self-shielded FCAW electrodes. Gas-shielded electrodes have greater depth of penetration.

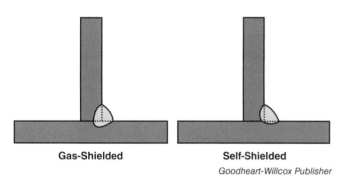

Goodheart-Willcox Publisher

Figure 13-8. Gas-shielded fillet welds in the flat and horizontal positions can be made smaller than the self-shielded type since penetration into the corner is deeper.

Electrode Selection

Consider the following factors when selecting an electrode wire:

- The mechanical and physical properties needed for the base material to be welded.
- The type and diameter of electrode needed for the joint.
- The position in which the welding will be performed.
- Whether single- or multiple-pass welding will be performed.
- Whether the electrode should be gas-shielded or self-shielded.
- The difference in cost between various types and sizes of electrodes based on deposition rates.

Equipment considerations include the following:

- Proper type of welding equipment for the intended electrode.
- Welding machine amperage capacity, including duty cycle.
- Type of gun, liners, contact tips, and nozzles.
- Wire feeder type, drive roller configuration, and size.

Shielding Gas

The electrode manufacturer specifies the proper shielding gas for a particular electrode. Always use the recommended gas or gas mixture unless you have specific instructions to do otherwise. FCAW-G requires either 100% CO_2 or a 75%–80% argon/ 20%–25% CO_2 shielding gas. Sufficient gas flow over the molten pool is needed to prevent air from entering and to enable the arc to melt off the wire in the proper pattern. Wind drafts disturb normal gas flow and may create problems with the arc pattern, leading to porosity in the weld.

Gas nozzles should be large enough to cover the molten weld pool and distribute the flow of gas over the entire weld width. During welding, remove spatter from the gas nozzle to allow proper gas flow. Antispatter spray and gelatin compounds can be applied to the torch nozzle and workpiece any time during the welding operation. Apply gel when the nozzle is warm so the gel melts and flows over the end

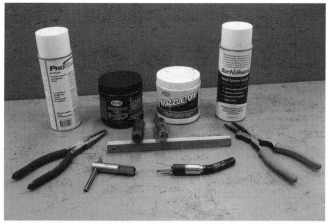

Goodheart-Willcox Publisher

Figure 13-9. Antispatter compounds and specialized nozzle cleaning tools are used to prevent the buildup of spatter in and on the nozzle tip.

of the nozzle. Specialized tools and pliers are used to remove built-up weld spatter. See **Figure 13-9**.

Test Welds

Established welding procedures for the carbon steels are listed in the *Reference Section* of this textbook. Wire type and size, amperage, polarity, stickout, welding position, joint design, and other variables must be considered before starting the actual welding operation. Keep the following tips in mind when making a test weld:

- Welds should be free of visible defects such as cracks, porosity, and undercut. If defects are present, change the welding technique or procedure before proceeding.
- When making multiple-pass welds, chip slag from the weld and clean any residue from the joint with a wire brush between each pass.
- Establish gun angles during the test to ensure penetration and proper bead contour.
- Stringer beads require less skill on the part of the welder. Weave beads create more distortion of the weldment.

- To control amperage and voltage, maintain an even electrode stickout during the actual welding operation. Doing so provides even penetration and proper metal deposition.
- When using a gas-shielded electrode, maintain the gun angle for even gas flow over the molten metal. When using CO_2 as the shielding gas, make sure the regulator does not freeze up and restrict gas flow. With high flow rates, use an electrical regulator with a heater system to avoid freeze-up.
- During welding, check the wire feeder for proper operation.
- If the power source is used for a long period of time, make sure it is within duty-cycle limits.
- The contact tip is a critical part in the system. If the tip is worn, the electrical current cannot properly contact the electrode. An interruption in current flow occurs, causing defective welds. Check the contact tip often and replace it when required.
- Gas nozzles attract spatter during welding. Use an antispatter compound and clean the nozzle often.

Test welds can prevent many problems. Not only can they be used for quality assessment, they also serve as a trial-and-error period for the welder. Test welds allow the welder to fine-tune the welding procedure, adapt new techniques for handling the gun, and observe the arc and bead pattern. Always record the welding parameters used on the test joint. If the test joint passes inspection, the recorded procedure can then confidently be used on the actual weld. If changes are made to the procedure, another test joint must be made to verify quality and required values.

In addition to serving as training and an opportunity to develop accurate procedures, weld tests help lower fabrication costs and encourage shop safety. Welding equipment and related apparatus, such as cables and the ground connection, should be checked for proper operation during a weld test. Know the condition of your equipment, and keep it in good repair.

Summary

- FCAW can be used to weld various types of carbon steels and alloys.
- FCAW electrodes for welding carbon steels and alloys are identified in AWS specification A5.36. Each electrode is identified by a series of letters and numbers that corresponds to the strength, chemical composition, and usability of the electrode.
- The two basic types of FCAW electrode wires are gas-shielded and self-shielded. FCAW electrodes are designed for particular types of metal deposition. Specific types of electrodes are manufactured for welding different types of alloy steels.
- Factors to consider when selecting an FCAW electrode include type of process, wire size, single- or multiple-pass weld, joint type and position, mechanical and physical properties, deposition rates, and type of equipment available.
- FCAW-G requires either 100% CO_2 or a 75%–80% argon/30%–25% CO_2 shielding gas.
- Wire type and size, amperage, polarity, stickout, welding position, joint design, and other factors must be considered before starting the actual FCAW operation.
- Test welds are used for quality assessment and also allow the welder to fine-tune the welding procedure, adapt new techniques for handling the gun, and observe the arc and bead pattern.

Review Questions

Answer the following questions using the information provided in this chapter.

1. *True or False?* Chrome-moly steels can be welded with the FCAW process.
2. An E71T-1 electrode can be used with which two metal deposition modes?
3. The type of electrode must be matched to the base material to obtain the desired _____ and _____ properties.
4. When is preheat, interpass temperature, and postheat important in the welding of low-carbon and mild steels?
5. Gas-shielded electrodes obtain better _____ because they result in higher current densities.
6. A fillet weld made with gas shielding can be made _____ but still have the same strength as a self-shielded weld.
7. What two types of shielding gas can be used with FCAW-G?
8. During welding, _____ should be removed from the gas nozzle to allow proper gas flow.
9. List three defects that require a change in welding procedure if they are observed during a test weld.
10. Why is it important to replace a worn contact tip?

Chapter

Flux Cored Arc Welding Procedures for Stainless Steels

Objectives

After studying this chapter, you will be able to:

❏ Select the appropriate FCAW electrode wire based on usability and performance capabilities.

❏ Prepare a stainless steel joint for welding.

❏ Determine the welding procedure required to produce a quality weld.

❏ Make a test weld on stainless steel using the FCAW process.

Technical Terms

carbide precipitation
run-off tab
run-on tab
stainless steels

FCAW Stainless Steels

Stainless steels are commonly welded with the FCAW process in the fabrication field. A small number of filler materials are used for stainless steels. Other types of electrodes are available for stainless steel alloys that are used for overlays and for joining dissimilar metals. Refer to Chapter 9 for detailed information about the various types of stainless steel and general information about welding stainless steels.

Stainless steels are sometimes called chromium-nickel steels because they have a high alloy content of chromium and nickel. These steels resist almost all forms of rust and corrosion. Stainless steels are iron-based alloys that contain a minimum of 10.5% chromium. These steels play an important role in the food, drug, and chemical refining industries. Stainless steels are used in conditions of extreme heat or cold. They cannot be hardened by heat treatment.

The American Iron and Steel Institute (AISI) identifies stainless steel by a numbering system and specifies the composition of each type. See **Figure 14-1**.

Electrode Wire

FCAW electrodes for stainless steels are listed in AWS A5.22 *Specification for Stainless Steel Electrodes for FCAW and Stainless Steel Flux Cored Rods for GTAW*. Unlike mild- and low-alloy steel FCAW electrodes, stainless steel electrodes are classified by material type rather than tensile strength. See **Figure 14-2** and **Figure 14-3**. When the word *tensile* appears in the specification, it refers to one of the mechanical properties of the material.

Commercially Wrought Stainless Steel Identification (AISI)

Composition Percent[a]

Type	Carbon (C)	Manganese (Mn)	Silicon (Si)	Chromium (Cr)	Nickel (Ni)	Phosphorus (P)	Sulfur (S)	Others
201	0.15	5.5–7.5	1.00	16.0–18.0	3.5–5.5	0.06	0.03	0.25 N
202	0.15	7.5–10.0	1.00	17.0–19.0	4.0–6.0	0.06	0.03	0.25 N
301	0.15	2.00	1.00	16.0–18.0	6.0–8.0	0.045	0.03	
302	0.15	2.00	1.00	17.0–19.0	8.0–10.0	0.045	0.03	
302B	0.15	2.00	2.0–3.0	17.0–19.0	8.0–10.0	0.045	0.03	
303	0.15	2.00	1.00	17.0–19.0	8.0–10.0	0.20	0.15 min.	0–0.6 Mo
303Se	0.15	2.00	1.00	17.0–19.0	8.0–10.0	0.20	0.06	0.15 Se min.
304	0.08	2.00	1.00	18.0–20.0	8.0–10.5	0.045	0.03	
304L	0.03	2.00	1.00	18.0–20.0	8.0–12.0	0.045	0.03	
305	0.12	2.00	1.00	17.0–19.0	10.5–13.0	0.045	0.03	
308	0.08	2.00	1.00	19.0–21.0	10.0–12.0	0.045	0.03	
309	0.20	2.00	1.00	22.0–24.0	12.0–15.0	0.045	0.03	
309S	0.08	2.00	1.00	22.0–24.0	12.0–15.0	0.045	0.03	
310	0.25	2.00	1.50	24.0–26.0	19.0–22.0	0.045	0.03	
310S	0.08	2.00	1.50	24.0–26.0	19.0–22.0	0.045	0.03	
314	0.25	2.00	1.50–3.0	23.0–26.0	19.0–22.0	0.045	0.03	2.0–3.0 Mo
316	0.08	2.00	1.00	16.0–18.0	10.0–14.0	0.045	0.03	2.0–3.0 Mo
316L	0.03	2.00	1.00	16.0–18.0	10.0–14.0	0.045	0.03	3.0–4.0 Mo
317	0.08	2.00	1.00	18.0–20.0	11.0–15.0	0.045	0.03	3.0–4.0 Mo
317L	0.03	2.00	1.00	18.0–20.0	11.0–15.0	0.045	0.03	$5 \times \%C$ Ti min.
321	0.08	2.00	1.00	17.0–19.0	9.0–12.0	0.045	0.03	1.0–2.0 Mo
329	0.10	2.00	1.00	25.0–30.0	3.0–6.0	0.045	0.03	
330	0.08	2.00	0.75–1.5	17.0–20.0	34.0–37.0	0.04	0.03	C
347	0.08	2.00	1.00	17.0–19.0	9.0–13.0	0.045	0.03	0.2 Cu[b, c]
348	0.08	2.00	1.00	17.0–19.0	9.0–13.0	0.045	0.03	
384	0.08	2.00	1.00	15.0–17.0	17.0–19.0	0.045	0.03	

a. Single values are maximum unless indicated otherwise. b. (Cb + Ta) min – 10 × %C. c. Ta – 0.10% max.

Goodheart-Willcox Publisher

Figure 14-1. The type numbers and compositions of common stainless steel base materials.

Stainless Steel Electrode Classification

EXXXT-X

E	Indicates an electrode.
XXX	Designates classification according to composition.
T	Designates a flux cored electrode.
X	Designates the external shielding medium to be used during welding.

Goodheart-Willcox Publisher

Figure 14-2. Stainless steel FCAW electrode classification.

Stainless Steel Electrode Types

E308LT-1	(Low-carbon material)
E309LT-1	(Low-carbon material)
E310T-1	
E312T-1	
E316LT-1	(Low-carbon material)
E317LT-1	(Low-carbon material)
E347T-1	

Goodheart-Willcox Publisher

Figure 14-3. Types of FCAW stainless steel electrodes.

All FCAW stainless steel electrodes operate on DCEP. Most use a 100% CO_2 or a 75% Ar/25% CO_2 shielding gas mixture.

Electrode Characteristics

Various companies manufacture FCAW electrode wire. Not every manufacturer makes every type of electrode. Characteristics of the electrodes available for stainless steels are shown in **Figure 14-4**. Class 1 and Class 3 electrodes deposit metal with the same basic chemistry. Class 1 electrodes transfer more of the required elements because the shielding gas provides a pathway for the molten metal and prevents air from entering the welding area. If chemical or physical tests are required, weld tests must be completed to determine the proper electrode, shielding gas, and welding procedure.

The American Welding Society requires tests on specific base materials for each type and lot of electrode. The deposited metal is then tested and must meet specific chemical and mechanical requirements to determine compliance with specification AWS A5.22. Chemical and physical test reports are available from electrode manufacturers for every lot of electrode. Individual spools or coils of electrode have tags identifying the lot.

Material Preparation

Stainless steel has a corrosion-resistant film on its surface caused by a reaction between the chromium/nickel and the atmosphere. For the most part, the film does not interfere with the quality of the completed

Stainless Steel FCAW Electrode Characteristics	
Class 1	**Electrode Characteristics**
E308LT-1	DCEP. CO_2 or Ar/CO_2 shielding gas. For stainless steel (S/S) types 301, 302, 304, 304L, 321, and 347.
E309LT-1	DCEP. CO_2 or Ar/CO_2 shielding gas. For S/S type 309 in cast or wrought form and type 304 to mild steel.
E310T-1	DCEP. CO_2 or Ar/CO_2 shielding gas. For S/S type 310 in cast or wrought form and dissimilar metals.
E312T-1	DCEP. CO_2 or Ar/CO_2 shielding gas. For joining dissimilar steels.
E316LT-1	DCEP. CO_2 or Ar/CO_2 shielding gas. For S/S type 316 with low carbon to prevent carbide precipitation and intergranular corrosion.
E317LT-1	DCEP. CO_2 or Ar/CO_2 shielding gas. For stainless steels with molybdenum for corrosion resistance.
E347T-1	DCEP. CO_2 or Ar/CO_2 shielding gas. For columbium stabilized S/S types 321 and 347.
Class 2	**Electrode Characteristics**
EXXXT-2	Ar/O_2 shielding gas that provides a spray transfer pattern of molten metal. No electrode wire currently available in Class 2 category.
Class 3	**Electrode Characteristics**
EXXXT-3	DCEP in all positions. Self shielded—no shielding gas required. Several electrodes available in Class 3 category.

Goodheart-Willcox Publisher

Figure 14-4. Characteristics of stainless steel FCAW electrodes.

weld. If any corrosion is present, remove it by grinding or brushing before welding. Do not use carbon steel wire brushes on the base metal or the weld. Always use stainless steel material brushes. If carbon steel brushes are used on the base metal before welding, carbon will be absorbed into the molten metal and could lead to *carbide precipitation* and intergranular corrosion. Carbide precipitation is a condition in which chromium leaves the grains and combines with carbon in the grain boundary.

Joints prepared by plasma arc cutting should be ground to bright metal before welding. Otherwise, oxides from the cutting are absorbed into the weld and cause defects. Oil or grease on the surface of the material should be removed before welding, as well.

Welding Procedures

Welding procedures for stainless steels are almost the same as for carbon steels. The major difference is the diameter of the welding electrode used for welding stainless steels. Small-diameter electrodes are used to minimize heat input where carbide precipitation is a problem and in overhead welding, where large molten pools cause metal sagging and excessive crown height.

Use *run-on tabs* and *run-off tabs* on long-seam welds to obtain penetration at the start of the weld and to prevent craters at the end of the weld. Run-on and run-off tabs are added pieces of base metal that are tacked onto the joint where the weld can be started and finished. These tabs are removed after the weld is completed. See **Figure 14-5**. Craters do not have enough weld cross section, so cracks form in these areas. A small tack weld at the seam intersection keeps the tabs in place during the welding operation. If tabs cannot be used, reverse the direction of the arc when you stop welding to build up metal and prevent craters.

Stainless steels retain heat much longer than carbon steels, so tooling may be necessary to remove heat from the weld area immediately after welding. Do not use carbon steel tooling because carbon transfers to the stainless steel wherever contact is made between the two.

Remember, stainless steel does not typically rust. Rust on stainless steel is caused by contact with carbon steel. Rust must be removed from carbon steels with a cleaner or acid to prevent transference to stainless steels.

Carbon steel is magnetic. When carbon steel is being welded to stainless steel, arc blow (deflection of the intended arc pattern by magnetic fields) may result if the work clamp is not properly placed.

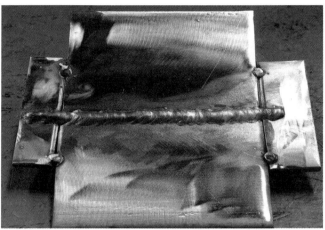

Goodheart-Willcox Publisher

Figure 14-5. The arc is started on the run-on tab. The weld is run from the run-on tab, along the entire joint, and ended on the run-off tab. The use of these tabs allows good penetration at the start and stop of the weld, without cold spots and craters.

Preheat, Interpass, and Postheat

Stainless steels generally do not require preheat, interpass, or postheat cycles. Problems are likely to occur when the base material gets hot. Stainless steels retain heat during welding, leading to cracks and carbide precipitation. If acid enters the grain boundary, it can cause intergranular corrosion or weld failure. To prevent cracking, use a low-carbon or extra-low-carbon filler material. Keep the weld heat to a minimum. Use stringer beads at a maximum interpass temperature of 300°F (149°C). The material should not be welded below 60°F (16°C). Stainless steels do not require postheating.

Joint Preparation

Joints for stainless steels are prepared similarly to joints for mild-steel materials, with the following exceptions:
- Self-shielded electrodes produce a slightly sluggish weld that may require more groove width and a larger bevel angle for manipulation of the molten metal. Joint groove widths and bevel angles should be fit up on the high side of the joint tolerance.
- Gas-shielded electrodes require a gas nozzle on the end of the welding gun. The joint opening should be sufficiently sized to make the joint and electrode extension visible so the operator can manipulate the gun.

Electrode Selection

Consider the following when selecting electrode wire:

- The electrode should closely match the composition of the base material. Always use a low-carbon electrode for low-carbon material. If the completed weld must have specific mechanical properties, check the test report for verification.
- Select the proper electrode diameter for the type of joint to be welded.
- Select the appropriate electrode type and diameter for the position in which the welding will be performed.
- Determine whether a self-shielded or gas-shielded electrode is needed.
- Review the deposition rates to compare the deposition costs of various sizes of electrodes.

Once the filler material is selected, coordinate the welding equipment. Considerations include the following:

- Machine capacity.
- Duty cycle.
- Type of gun and nozzles.
- Contact tips.
- Cable liners. Always use nylon or Teflon™ liners with stainless steel electrodes. Check liners often during the operation and replace them when needed.
- Wire feeder drive roll size and design. Drive rolls may be either *V* or *U* type. Use only finger-tight feed-roll pressure because too much pressure causes the electrode to crack and the liners to wear excessively.
- Type of shielding gas.

Shielding Gas

The proper shielding gas for a particular electrode is specified by the electrode manufacturer. Using a mixture other than the one specified could change the chemical composition of the weld. When chemical requirements are imposed, perform a weld test to verify that the final chemical composition complies with the specification. Never use shielding gas on an electrode designated as self-shielded.

Sufficient gas flow is required over the molten pool to prevent air from entering the weld. Wind and drafts disturb normal gas flow and create arc pattern problems, causing porosity in the weld. Flow rates for gas-shielded welds vary from 25 cfh to 35 cfh, depending on wind gusts in the weld area.

Gas nozzles should be large enough to distribute the shielding gas over the molten pool. Any spatter that accumulates in the nozzle should be removed and an antispatter compound applied to the interior of the nozzle.

Tooling and Backing

Nonmagnetic tooling must be used to prevent magnetic arc blow. This type of tooling also prevents iron pickup on the top and back of the base material. A shielding gas backing purge should be used to prevent oxidation on full-penetration welds.

Where gas backing cannot be used, a flux substance can substitute. The paste is mixed with alcohol, applied to the part, and allowed to dry before welding. After welding, the flux is removed with warm water and a stiff stainless steel brush. Further cleaning of the stainless steel can be done with a commercial stainless steel cleaner.

Test Welds

As with carbon and low-alloy steels, performing test welds on stainless steels prevents many problems. The test welds can be used for quality assessment and serve as a learning period for the welder. Ultimately, test welds confirm that the welding procedure is satisfactory and the weld can be properly made. The welder has the opportunity to fine-tune the welding procedure, adapt new techniques for handling the gun, and observe the arc and bead pattern. Welding equipment should be checked for proper operation during this period. Check contact tips often for signs of wear.

The *Reference Section* contains several procedures for welding stainless steels with different types and sizes of electrodes. These procedures are general and should be verified before being used for production.

Testing and Chemical Analysis

Problems can and do occur in the welding of stainless steels. Many specifications and standards for welding require special tests for weld qualification. Others require chemical analysis of the actual deposited weld metal to ensure quality. These measures assure the purchaser that the weldment will operate satisfactorily in the environment for which it was made. If a particular procedure requires special tests or chemical analysis, investigate the welding areas with a qualified metallurgist before ordering welding material and consumables, designing weld joints, and developing welding procedures.

Summary

- Stainless steels are commonly welded using both gas-shielded and self-shielded FCAW processes.
- The electrode wire should be determined based on the type of stainless steel and should match the base metal as closely as possible.
- Stainless steels generally do not require preheat, interpass, or postheat.
- Welding procedures for stainless steels are almost the same as for carbon steels. The major difference is that when stainless steels are welded, a smaller-diameter FCAW electrode is used to minimize heat input.
- A stainless steel wire brush and designated grinding disks should be used to prepare and clean the base metal.

Review Questions

Answer the following questions using the information provided in this chapter.

1. Stainless steels contain a minimum of 10.5% _____.
2. *True or False?* Stainless steel electrodes are classified by tensile strength.
3. Tests on specific base materials for each type and lot of electrode are required by the _____.
4. Explain why carbon steel wire brushes should never be used to remove corrosion from stainless steels before welding.
5. A groove weld made with a self-shielded electrode requires a larger _____ for the molten metal to flow properly.
6. Why should joints prepared by plasma arc cutting be ground to bright metal before welding is begun?
7. Why are run-on tabs and run-off tabs used on long-seam welds?
8. *True or False?* Stainless steels generally require postheating.
9. What type of filler material should be used when stainless steel is welded in order to prevent cracking?
10. A shielding gas backing purge or flux should be used to prevent _____ on full-penetration welds.

Chapter

Flux Cored Arc Welding Procedures for Cast Iron

Objectives

After studying this chapter, you will be able to:

❑ Describe the four types of cast iron.

❑ Select the proper electrode based on usability and performance capabilities.

❑ Prepare cast iron for welding.

❑ Establish an FCAW welding procedure for cast iron.

Technical Terms

cast irons
ductile iron
ductility
gray iron
malleable iron
white iron

Cast Irons

Cast irons contain 93% to 95% iron, 1.75% to 4.5% carbon, and smaller quantities of other elements, such as chromium, nickel, manganese, and copper. The elements vary depending on the type of casting, heat treatment, and final use. Basic composition requirements for cast iron materials are as follows:

- Iron, 93%–95%.
- Carbon, 1.75%–4.5%.
- Silicon, 0.5%–3.0%.
- Manganese, 0.5%–1.0%.
- Phosphorus, 0.15%–1.0%.
- Sulfur, 0.05% or more.

As the name *cast iron* implies, the material is melted and cast (poured into a mold) to shape the desired part. Carbon gives cast iron very high strength and very low *ductility* (ability to deform without breaking). During the initial cooling and solidification, the carbon takes on new forms that determine the final characteristics of the material.

Types of Cast Irons

There are four types of cast irons:

- Gray iron
- White iron
- Malleable iron
- Ductile or nodular iron

Gray iron is the most commonly used form of cast iron. It casts easily in sand molds and is inexpensive to produce. Heat treatment can be done after casting to improve the mechanical properties. When gray cast iron is cooled, the carbon forms flakes of graphite. The edges of the graphite allow cracks to form when the material is under stress. A cracked surface appears gray, hence the name *gray iron*. The material has very low ductility. Some alloys of this type of cast iron are readily welded with the FCAW process using specific welding procedures.

White iron is rapidly quenched in a water-cooled mold during casting to prevent formation of graphite in the grain structure. However, graphite does form on the casting surface, which becomes hard and abrasive. The basic casting has very low ductility; therefore, white iron is not considered weldable.

Malleable iron is made by heat-treating white iron to change the graphite flakes to a spheroidal shape. This change reduces the tensile strength of the material but improves the ductility. Some of these alloys are readily weldable with the FCAW process.

Ductile iron is also called *nodular iron* or *spheroidal iron*. During the initial melting of the ore, various elements are added to create spheres of the excess carbon, resulting in an extremely strong and ductile material. Adding alloys and varying the heat-treating cycles results in various tensile strengths. Higher tensile strengths are directly affected by welding, which can result in the formation of cracks and hard heat-affected zones. All grades of ductile

cast iron are weldable using FCAW with the proper welding procedure.

Filler Materials

Filler materials for welding cast irons include steel cored electrodes or an iron-nickel electrode developed for cast-iron applications. The steel electrodes meet specification AWS A5.20 for classifications EXXT-4, 6, 7, and 8.

The iron-nickel cored electrode is not listed under an FCAW electrode specification. However, the filler material requirements meet specification AWS A5.15 for classification ENiFe-C1.

Electrode Characteristics

The previously mentioned steel electrodes do not match any of the chemical requirements of cast iron. Therefore, their use is limited to joining cast iron to steel or to making noncritical repairs. Because of its mechanical properties, the iron-nickel electrode is an exceptional choice for welding cast iron to cast iron, joining cast iron to other types of metals, and repairing cast iron work. See **Figure 15-1** for characteristics of carbon and low-alloy steel FCAW electrodes.

Carbon steel electrodes pick up carbon from the base metal, resulting in a weld that is high in tensile strength and low in ductility. Depending on the joint design, weld procedure, amount of preheat, and final

Carbon and Low-Alloy Steel FCAW Electrode Characteristics	
Steel Electrodes	**Characteristics**
EXXT-4	DCEP. Self-shielded. Globular transfer mode. Weld is hard and cannot be machined. Has little resistance to cracking.
EXXT-6	DCEP. CO_2 or 75% Ar/25% CO_2 shielding gas. Globular transfer mode. Low-hydrogen type flux. Produces a more crack-resistant weld than a T-4 electrode.
EXXT-7 **EXXT-8**	DCEN. Self-shielded. In small diameters, can be used in all positions. Welds are extremely hard, have very little crack resistance, and are nearly impossible to machine.
ENiFe-C1-A	DCEP. Shielding gas varies depending on the electrode supplier. Manufacturer may specify the use of 98% Ar/2% O_2, or no shielding gas required.

Goodheart-Willcox Publisher

Figure 15-1. Characteristics of carbon and low-alloy steel electrodes used in FCAW welding of cast iron.

heat treatment of the completed weld, the welded joint should have sufficient strength. However, the low ductility of the weld may cause failure of the joint when it is placed under stress.

The iron-nickel-manganese electrode (ENiFe-C1-A) consists of 45% iron, 50% nickel, 1% carbon, and approximately 4% manganese. It is specially produced for welding cast irons. Tensile strength is not a problem with this electrode. The ductility of the weld and heat-affected zones are improved by the nickel content and the addition of carbon to the weld zone.

Material Preparation

Clean the weld area and remove any dirt, rust, scale, or grease before the welding operation. Such contaminants affect the deposition of the weld metal and may cause defects in the weld or weld interface. Castings with surface scale should be ground to remove the scale. Follow safety instructions and always wear eye protection when grinding. Thermal-cut materials have oxidized metal on the face of the cut, providing a place for weld defects to form. Grind these areas to bare metal before welding.

Grease burns and leaves a residue that causes porosity in the weld. Use degreasers to remove grease from the weld area. Follow recommended safety rules when using degreasers.

Cracks in castings are caused by stresses beyond the strength of the material. The end of the crack may not be visible but may be found with a liquid penetrant test, which is described in Chapter 17. Ensure that dye and developers are completely removed from the crack before welding. Often, holes are drilled at each end of the crack to prevent further cracking during the heating, welding, and cooling operations.

Welding Procedures

Cast iron is melted, poured into a mold, and slowly cooled. Slow cooling allows stresses to be relieved and prevents cracks from forming. Regardless of the technique used, welding stresses the casting. Therefore, the procedure must almost duplicate the original casting procedure in terms of solidification of the metal and cooling of the casting.

Basic welding operations include the following:
- Repair of surface defects in the shell of the casting.
- Addition of sections to complete an assembly design.
- Repair of a worn area or break in the casting.

Preheat, Interpass, and Postheat

Preheating, interpass heating, and postheating may or may not be critical in welding a casting. Thin parts or attached components made of other types of materials may not need preheating. However, a large casting with a thin part attached may require heating the large part to prevent cracking. A shallow skin defect may not need preheating; however, preheat may be required if this defect is located in a complex design area.

For the most part, studying the weld to be made and using common sense will help you determine whether to apply heat and for how long. Remember that material expands when heated and contracts (shrinks) when cooled. If the material cannot contract on cooling, stresses form and cracks result.

Preheating heats a local area or an entire part. For small local areas on the casting skin, preheat is not required unless moisture is present. Moisture can also be removed with alcohol, acetone, or air. If the weld is in a complex, stressed area, full preheat is probably required. Full preheat ranges from 600°F (316°C) to 1000°F (538°C) for high-carbon material.

During welding, the interpass temperature of the entire part must not fall below the preheat temperature. If the part is postheated or stress-relieved after being welded, the casting may crack when the part is placed in the furnace.

Postheat is just as important as preheat. Postheating must be completed before the part is allowed to cool below the interpass temperature. Heating blankets should be applied to the part to slow cooling, thereby eliminating stresses and potential cracking.

Whenever any type of heating is used, the part must be protected to prevent rapid cooling. Windy areas can cause cooling and cracking in the weld or heat-affected zone.

Joint Preparation

When designing a joint, do the following:
- Design the joint to compress during welding, **Figure 15-2**.
- Be aware that single V-groove welds shrink more than double V-groove welds, **Figure 15-3**.
- Design weld joints of equal thickness, **Figure 15-4**.
- Use fillet welds for minimum shrinkage.
- Butter component parts or joint sidewalls to reduce the amount of carbon pickup from the groove wall and to reduce the hardness of the weld. See **Figure 15-5**.
- Replace broken pieces with steel materials, if possible. Allow for weld shrinkage. See **Figure 15-6**.

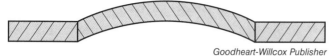

Goodheart-Willcox Publisher

Figure 15-2. A domed piece of material shrinks during welding, reducing the possibility of the base material cracking.

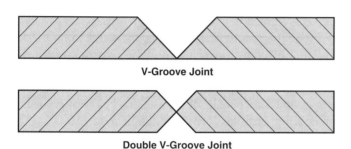

V-Groove Joint

Double V-Groove Joint

Goodheart-Willcox Publisher

Figure 15-3. A V-groove joint requires twice as much filler material as a double V-groove joint.

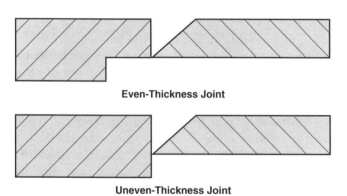

Even-Thickness Joint

Uneven-Thickness Joint

Goodheart-Willcox Publisher

Figure 15-4. Even-thickness joints allow the material to shrink evenly. Uneven-thickness joints do not cool at the same rate, causing stresses to build up in the joint and possibly resulting in weld failure.

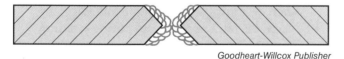

Goodheart-Willcox Publisher

Figure 15-5. Buttering welds reduces the possibility of shrinkage cracks because the filler material has more elongation than the cast iron.

Test Welds

The welding operation depends on a correct procedure for the deposition of the filler material. Typical welding parameters for various diameters of electrodes are listed in **Figure 15-7.**

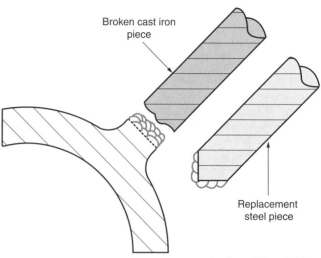

Broken cast iron piece

Replacement steel piece

Goodheart-Willcox Publisher

Figure 15-6. Grind the break area to a clean, shiny finish, inspect for additional cracks, and then butter the end of the break with several layers using a nickel-based electrode wire. The replacement piece can also be built up with the same nickel-based electrode wire, if desired. Complete the weld using this electrode.

Welding Parameters			
Alloy Rods Corp. NICORE 55 electrode. Direct current, reverse polarity (DCRP); 98% Ar/2% O_2 shielding gas at 20–45 cfh.			
Diameter	**Amps**	**Volts**	**Stickout**
0.035″	150–180	26–28	1/2″
0.045″	220–250	27–29	1/2″
1/16″	280–320	28–30	5/8″
3/32″	380–420	30–32	5/8″
Huntington Alloys NI-ROD FC 55 electrode. DCRP; CO_2 shielding gas at 30 cfh.			
Diameter	**Amps**	**Volts**	**Stickout**
0.078″	310	29	3/4″
0.093″	350	29	3/4″

Goodheart-Willcox Publisher

Figure 15-7. These are general welding parameters. Consult with the filler material manufacturer for specific information.

Set your machine to produce sufficient amperage and voltage consistent with good fusion. Establish the machine setting by welding on a piece of scrap metal and observing metal deposition. If you intend to preheat the actual weld, preheat the test part as well. The required welding current will be lower for a preheated part than for a cold part. Once you have established machine parameters and variables, record the settings for future use. These settings will serve as a base procedure from which changes can be made.

Welding Tips for FCAW of Cast Irons

When using the FCAW process, welders tend to make longer welds than when using other processes. Follow these suggestions to help eliminate problems encountered when cast iron is welded with the FCAW process:

- Remove porosity in the weld by grinding or chipping, *not* by welding over it. When a weld is done, peen it immediately with a rounded tool. Use rapid, moderate blows. This expands the weld metal and reduces the amount of weld shrinkage.
- To prevent slag from moving ahead of the weld, slightly tilt the gun angle backhand. Always remove slag and wire-brush a completed weld before depositing another layer of material.
- Stringer beads work best because the heat from the arc is applied to a small area. Wide wash beads apply heat over a greater area, resulting in greater weld shrinkage and distortion, leading to cracking in the heat-affected zone.
- Backstep welding and skip welding help reduce heat input and cracking at the weld interface and heat-affected zones, **Figure 15-8**. Make short welds so you can peen the completed weld before it cools.
- Do not try to hurry the welding. Stay with your procedure.
- Inspect each weld bead for cracks, lack of fusion, and porosity. Adjust the parameters to eliminate these conditions. The weld bead should not have a high crown. A high crown on the weld can cause cracking at the weld interface when stressed. Remove these areas and any undercut before allowing the casting to cool. If air-powered tools are used, do not allow the exhaust

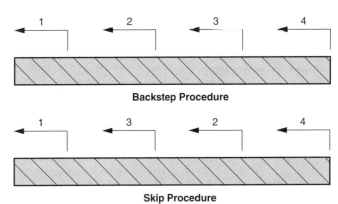

Backstep Procedure

Skip Procedure

Goodheart-Willcox Publisher

Figure 15-8. Backstep and skip welding reduce expansion and subsequent joint shrinkage during welding. These techniques are often used when the part cannot be heated to a higher temperature.

to flow onto the casting. Rapid chilling in a local area may cause cracking.
- If the procedure requires postheating, do so immediately. If a furnace operation is required for stress-relief of the weld, make sure the part is at an even temperature *before* it enters the furnace.

In addition to the previous suggestions, keep the following points in mind:

- High voltage causes air contamination of the arc column and potential porosity of the weld.
- High voltage causes spatter.
- High current increases penetration and increases the mixture of the filler material and base material.
- Decreasing current reduces undercut.
- Decreasing stickout increases penetration.
- Increasing stickout can overheat the electrode and cause spatter.
- Stringer beads reduce shrinkage.
- Weave beads reduce porosity of the weld.
- Weave beads, since they are wider than stringer beads, require longer welding times for a particular length of weld. This increases the amount of shrinkage as the weld cools, increasing the potential for shrinkage cracking.

Summary

- Cast irons contain 93% to 95% iron, 1.75% to 4.5% carbon, and smaller quantities of other elements. The material is melted and cast (poured into a mold) to shape the desired part.
- Carbon gives cast iron very high strength and very low ductility.
- The four types of cast irons are gray iron, white iron, malleable iron, and ductile (nodular) iron.
- The iron-nickel electrode (ENiFe-C1) is an exceptional choice for welding cast iron to cast iron, joining cast iron to other types of metals, and repairing cast iron work.
- Basic welding operations include repair of surface defects in the shell of the casting, adding sections to complete an assembly design, and repair of a worn area or break in the casting.
- During welding, the interpass temperature of the entire part must not fall below the preheat temperature. Postheat is just as important as preheat. Whenever any type of heating is used, the part must be protected to prevent rapid cooling.

Review Questions

Answer the following questions using the information provided in this chapter.

1. The carbon in cast iron gives it very high _____ and very low _____.

2. Small flakes of _____ form during the solidification of gray iron in the cooling mold.

3. The spheroidal graphite flakes in malleable iron result in reduced _____ and improved _____.

4. List two other names for ductile iron.

5. List three types of cast-iron welding applications for which the iron-nickel electrode can be used.

6. When cast iron is welded with an ENiFe-C1-A electrode, the _____ of the weld and heat-affected zones are improved by the nickel content and the addition of carbon to the weld zone.

7. Materials that have been thermal-cut should always be ground to _____ prior to welding.

8. Preheating of small local areas on the casting skin is not required unless _____ is present.

9. *True or False?* During welding, the interpass temperature of the entire part should fall below the preheat temperature.

10. *True or False?* Single V-groove welds shrink more than double V-groove welds.

11. Buttering component parts reduces the amount of _____ from the groove wall and reduces the hardness of the weld.

12. Why should a finished weld be peened immediately, before it cools?

13. What two intermittent weld techniques help reduce heat input and cracking at the weld interface and heat-affected zones?

14. Why should the weld bead *not* have a high crown?

15. *True or False?* Decreasing stickout increases penetration.

Chapter 16

Flux Cored Arc Welding Procedures for Surfacing Welds

Objectives

After studying this chapter, you will be able to:

❏ Describe the functions of different types of surfacing.

❏ Distinguish between different types of electrode wires used for surfacing.

❏ Prepare material for surfacing.

❏ Describe procedures for making surfacing welds.

❏ Determine whether surfacing is cost-effective.

Technical Terms

buildup welds
buttering
cladding
hardfacing
stepover distance

Surfacing

Various metals are surfaced or covered with other metals to protect them from another material that could cause wear or deterioration of the original part. Wear includes abrasion, impact, heat, corrosion, or a combination of these factors. The part can be made undersize, surfaced, then machined to size as a new part, or it can be rebuilt from a used part. In any case, the surfacing operation extends the life of the part at considerable cost savings. *Hardfacing* is a special form of surfacing applied to make a part harder as a means of reducing wear. See **Figure 16-1**. In addition to hardfacing, there are several other types of surfacing operations. *Buttering* refers to surfacing an area where materials are to be joined and

Goodheart-Willcox Publisher

Figure 16-1. This tractor bucket plate was hardfaced for wear resistance.

a dissimilar material is used for the joining. This type of weld can also be used when two component parts are to be welded and heat-treated to move the joint stress from the components. Buttering is also used to obtain specific chemical values in the final weld. For example, when stainless steel is applied onto carbon steel and the final deposit needs considerably lower carbon content, a different stainless steel is used for buttering than is used for the final joining weld.

Cladding is the application of a material that provides a corrosion-resistant surface. Typical corrosion-resistant clad layers include copper-based weld overlays for saltwater resistance, nickel alloys for saltwater or sour gas service, and various Stellite® electrodes for corrosion and wear resistance. *Buildup welds* are made to restore worn parts to original dimensions or to create a base layer for the application of cladding or hardfacing materials. See **Figure 16-2**.

Base Materials

Cast irons, low- and high-alloy steels, stainless steels, and manganese steels are the materials most commonly surfaced with the FCAW process. The mining, steelmaking, railroad, and construction industries use many types of base materials for the manufacture of their products and equipment. Many materials have been developed or modified for use in a specific application.

Goodheart-Willcox Publisher

Figure 16-2. This large flanged connection tube is being built up by FCAW before being machined to the required thickness.

Electrode Wires

Some electrode wire fits into an AWS specification for a particular material. However, many of the surfacing materials have been developed by electrode manufacturers for specific applications and are not classified by AWS. A number or a trade name lists these proprietary electrodes.

An electrode may be designed for buildup, impact resistance, hardfacing, or resistance to corrosion and heat. The various electrodes and the chemical elements used in their manufacture are described next.

Buildup Electrodes

Buildup electrodes may or may not match the base metal's composition and the mechanical values established as a base for the final surfacing material.

Carbon-steel electrodes consist of the following elements:
- Iron-base core.
- Carbon.
- Manganese.
- Silicon.
- Chromium.
- Molybdenum.

Total alloy content of carbon-steel electrodes is low. The material has good impact strength and machinability, making it a suitable base for application of the final surfacing material. Carbon-steel electrodes are used for buildup that may require multiple layers. Buildup layers should be made within one or two layers of the final dimension. The final surfacing material is then applied. Carbon-steel electrodes are designed for welds on all weldable carbon and low-alloy steels. The welds have a high resistance to deformation.

Manganese-steel electrodes consist of the following elements:
- Iron base core.
- Carbon.
- Manganese.
- Nickel.

Manganese-steel electrodes can be used to join manganese steels, repair worn areas, and build up for surfacing. The deposit work hardens to a high tensile strength and has good impact properties.

Buildup electrodes for welding on cast iron are limited to the steel electrodes and the iron-nickel electrodes. The use of steel for a buildup electrode

Surfacing Electrode Composition	
Type of Electrode	**Application**
Stainless steel	Used as an overlay for the carbon steels for corrosion protection. Type 309L S/S is used for the first layer, and type 316L is used for the final surfacing weld.
High percentages of manganese and chromium	Used on materials requiring protection from moderate to severe abrasion. These electrodes generally work-harden in use and cannot be flame cut or machined after welding.
Nickel base with manganese, chromium, tungsten, and molybdenum	Used for resistance to high heat and corrosion, with good impact and abrasion values.
Iron base with small amounts of tungsten carbides within the core	Have excellent resistance to earth wear.

Goodheart-Willcox Publisher

Figure 16-3. Chemical composition of surfacing electrodes and their applications.

creates extreme hardness in the weld area due to carbon pickup from the cast iron. The material then cracks along or across the weld, leaving a poor base for applying the final surfacing material. Iron-nickel-manganese steel electrodes are strongly recommended for this application.

Surfacing Electrodes

Surfacing materials can usually be applied to any of the buildup materials and many base materials. The main objective is to retain the physical properties of the surfacing material after the welding operation. This is accomplished by using the correct procedures to minimize the depth of penetration on all buildup and surfacing welds and by properly adding the final surfacing material. Incorrect procedures or excessive dilution degrades the weld quality and lowers the surfacing protection provided. Contact the surfacing material manufacturer for the proper welding procedure.

During surfacing operations on construction equipment using extremely high-alloy-content electrodes, the weld may tend to crack either across or in line with the weld. Rapid cooling of the weld and extreme hardness of the deposited metal causes this cracking. In most cases, these cracks do not have an adverse effect on the usability of the completed weld.

Through experience and physical testing, each manufacturer has established the chemical values of surfacing electrodes and the procedures for applying them. The electrodes and chemical requirements in **Figure 16-3** are general and serve only as a guide for selection of the desired electrode. The manufacturer has thoroughly tested the welding material and found

the recommended electrode works well for the application indicated.

Material Preparation

Clean the weld area and remove any dirt, rust, scale, or grease before welding. Such contaminants affect the deposition of the weld metal and may cause defects in the weld or weld interface. Remove surface scale from castings by grinding. Thermal-cut materials have oxidized metal on the face of the cut, providing a place for weld defects to form. Grind these areas to bare metal before welding.

Grease burns and leaves a residue that causes porosity in the weld. Remove grease from the weld area with degreasers. Be sure to follow recommended safety rules when using degreasers. In areas where parts of previous weld deposits remain, remove the old material to bare, shiny metal. Inspect the material thoroughly and repair cracks before applying buildup or surfacing materials. Follow safety instructions and wear eye protection when grinding.

Welding Procedures

Each type of surfacing requires a different procedure, depending on the base material and type of welding electrode used. See **Figure 16-4**.

During surfacing, the material must be deposited into the base or buildup material with a minimum amount of dilution to maintain the proper metallurgical content of the filler material. The manufacturer of the filler material specifies the type of current and

Goodheart-Willcox Publisher

Figure 16-4. Button-type welds and a diamond pattern have been added to reduce wear on the blade.

shielding gas (if required). The amperage should be sufficient to maintain a steady arc and good pool control. The electrode should be almost perpendicular to the weld, with only a slight amount of drag angle, in order to reduce penetration. However, slag may flow forward and become entrapped if the arc is too long. High arc voltages and long electrode extension reduce penetration and minimize dilution, but they tend to cause lack of fusion and slag entrapment.

The amount of dilution into the base material has a direct effect on the chemical composition of the final pass. To establish the proper amount of dilution of the weld metal, a test can be made to determine the actual welding parameters and the final results of the weld. During the test period, all welding parameters must be recorded so the procedure can be duplicated for the actual weld. If the desired qualities are not achieved during the test period, another test must be made with another set of welding parameters.

Surfacing welds are best applied in the flat position to limit heat input and impart the best bead shape to the completed weld. After the first pass is made, the remaining passes are usually located on half of the completed weld and half of the base metal. The distance moved over to make this next weld is called

the *stepover distance*. This distance must be maintained until all the required welds are completed. Moving the gun across the weld path in a set pattern makes oscillated welds. These types of welds are made more slowly. More distortion occurs, and base material heat can affect the amount of dilution.

Oscillated welding is easily controlled on an automatic machine where the oscillation parameters can be determined and controlled. Welders using semiautomatic equipment should make test welds with the proposed procedure and production equipment to verify the welding parameters to be used.

Each weld of the overlay should be made to a specific procedure to maintain the chemical limits of the final weld. A weld that does not meet this requirement will fail in service, resulting in excessive repair costs or jeopardizing the safety of the equipment operator.

Preheat, Interpass, and Postheat

The temperatures for various types of surfacing operations depend on many factors, including the following:
- Type of base material.
- Thickness of the base material.
- Complexity of the weldment design and type of weld joint.
- Type of filler materials to be used.
- Mechanical and chemical values of the completed weld.
- Final heat treatment of the surfaced part.

Other areas must be considered before welding the part. Some metals cannot be heated beyond a certain temperature—the heat destroys the grain structure of the base material and causes the weldment to fail. Other metals must be postheated to prevent hardening of the final welds and cracking as the weld metal cools.

The following conditions for heating operations generally apply:
- Austenitic manganese steels are *not* heated.
- Stainless steels are *not* heated.
- Low-carbon steels are *not* heated.
- Cast irons are heated.
- Extremely cold materials are heated. (Thermal shock can cause cracking of the base material or the weld.)
- High-chromium steels are heated.
- Medium- and high-carbon steels are heated.
- Very hard surfacing deposits are heated.

In general, if the part is to be heated, the entire heat-affected zone requires heat. Complex design parts must be heated throughout to prevent cooling stresses from forming. If preheating is required, interpass heating and postheating are done also. Interpass temperature should be maintained until the welding operation is complete. The final heating should be started before the part drops below the interpass temperature. Final cooling should be done by covering the part with an insulating material or a blanket or placing it in a furnace to prevent rapid cooling. Rapid cooling results in internal or external stresses that may cause cracking of the weld or the weldment.

Hardfacing Patterns

Many completed hardfacing welds on construction equipment have patterns that cover only part of the wearing surface. See **Figure 16-5**. Labor and materials would be too costly to cover the entire surface area. The weld patterns act as dams, while the "open" areas between the weld passes fill with dirt or sand, which compact in the spaces between welds and protect the metal underneath from further abrasion.

All types of welding designs are used to protect the base material from wear, depending on the type of earth being moved. Designs include buttons, circles, curved lines, zigzagged lines, and straight lines with large and small gaps. To ensure that the welds effectively prevent wearing of the part, take into account the final use of the equipment when choosing a pattern.

For example, different patterns and materials are used for a bucket that primarily moves hard materials, such as gravel, and a bucket that primarily moves soft materials, like dirt or sand. The hardfacing for gravel and stone must provide protection against both abrasive and impact wear. A pattern that is parallel to material flow provides the best wear resistance for hard materials. A bucket that primarily moves soft materials like dirt or sand is exposed to little impact wear, but a great deal of abrasive wear. A pattern that is perpendicular to material flow provides the best wear resistance for soft materials. A waffle or diamond pattern is typically used on equipment that moves a combination of hard and soft materials. See **Figure 16-6**.

Goodheart-Willcox Publisher

Figure 16-6. The weld pattern traps dirt or sand in the pockets. The open areas are protected as if they were welded. Note the pattern used and the amount of the original area left uncovered.

Goodheart-Willcox Publisher

Figure 16-5. Many patterns were used to hardface this dragline bucket.

Goodheart-Willcox Publisher

Figure 16-7. Bucket housings for these front blade teeth were hardfaced.

Safety

Take safety precautions when surfacing to prevent accidents and injuries. Working around heavy machinery requires being alert to others working near you. Once your welding helmet is down, you cannot see your surroundings. Let others know you are there so they can help you, if needed. If possible, have a person work with you at all times. Make sure you have all the equipment necessary to complete the task. Keep a fire extinguisher handy when working near gasoline or diesel engines, fuel tanks, hydraulic systems, or other flammable materials.

Test Welds

The actual welding operation should match the test procedure as closely as possible. The test procedure determines the correct welding parameters and variables necessary to complete the task and obtain a satisfactory weld. The welder who develops the procedure should carry out the actual welding or should instruct the assigned welder in the proper procedure. A test weld should be made after the machine is set up to confirm the settings are correct. A change in electrode diameter, welding voltage, stickout, or travel speed can affect the deposition characteristics and cause the weld to fail.

Alternatives to Surfacing

Type of surfacing material, weld deposition time, and final use of the equipment are three factors that determine if a welding operation is cost-effective. If surfacing is too expensive, a new part may be used and the old, worn-out part thrown away. This is especially true in rework on tractors used for earthmoving. The front teeth on the bottom of the blade typically are not surfaced because the cost is too high. Instead, formed plates covering the entire part are applied. Welding is used only to attach the new plate to the old one. Sometimes a surfacing operation is not practical because of the design of the part. In these cases, wear plates are used and attached by welding, **Figure 16-7**.

Electrode and labor costs are high. However, filler materials for FCAW are considerably less expensive than filler materials for SMAW. When equipment is repaired in the field, the use of portable equipment reduces downtime and lowers costs for the operator.

Summary

- Surfacing reduces maintenance costs on equipment and is used to increase the useful life of new equipment before it is placed in service.
- Hardfacing is a special form of surfacing applied to make a part harder as a means of reducing wear. Not all surfacing welds are applied to make the part harder.
- Buttering is the process of surfacing an area where materials are to be joined and a dissimilar material is used for the joining. This type of weld is also used to move the joint stress away from the components when two component parts are to be welded and heat-treated.
- Cladding is the application of a material that provides a corrosion-resistant surface.
- Buildup welds are made to restore worn parts to original dimensions or to create a base layer for cladding or hardfacing materials. The two types of buildup electrode wire are carbon steel and manganese steel. The required results determine which electrode wire should be used.
- Each type of surfacing requires a different procedure, depending on the base material and type of welding electrode used.
- The object of surfacing is to deposit the material into the base or buildup material with a minimum amount of dilution to maintain the proper metallurgical content of the filler material.
- Take safety precautions when surfacing to prevent accidents and injuries.

Review Questions

Answer the following questions using the information provided in this chapter.

1. What is the purpose of surfacing?
2. List four causes of wear to metal parts.
3. Surfacing an area where materials are to be joined and a dissimilar material is used for the joining is referred to as _____.
4. What materials are most commonly surfaced with the FCAW process?
5. What two properties of carbon-steel electrodes make them a suitable base for application of the final surfacing material?
6. What two types of buildup electrodes are used for cast iron?
7. To retain the values of the surfacing material, the correct procedures must be used to _____ the depth of penetration on all buildup and surfacing welds.
8. Surface scale should be removed from castings by _____.
9. During surfacing, the material must be deposited into the base or buildup material with a minimum amount of _____ to maintain the proper metallurgical content of the filler material.
10. During surfacing, how should the electrode be held in order to reduce penetration?
11. How can the proper amount of dilution of the weld be established?
12. Why is it best to apply surfacing welds in the flat position?
13. Oscillated welds are made slowly. As a result, more _____ occurs, and base material heat can affect the amount of dilution.
14. Why should final cooling of a part be done slowly?
15. When weld patterns are used for hardfacing, the open areas between the weld passes fill with _____.

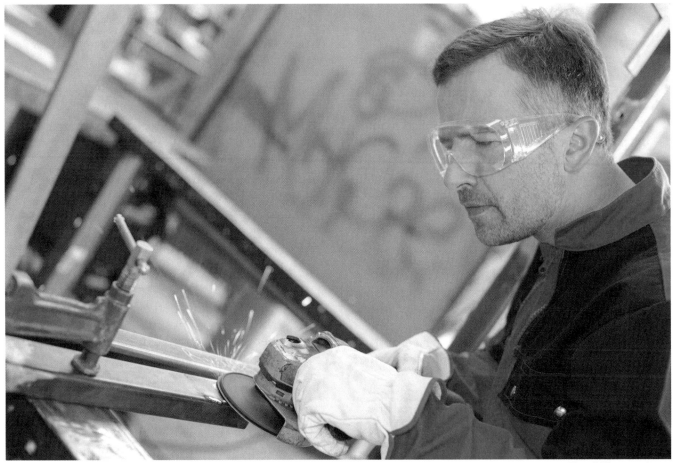

Follow proper safety procedures and wear eye protection when grinding.

Welding Procedures, Defects, and Corrective Actions

Objectives

After studying this chapter, you will be able to:

❏ Define a qualified welding procedure.

❏ Describe five nondestructive inspection methods.

❏ Identify and correct GMAW and FCAW defects in groove, fillet, plug, and spot welds.

Technical Terms

defects
destructive testing
discontinuities
ferromagnetic
liquid penetrant inspection
magnetic particle inspection
nondestructive testing
qualified procedure
radiographic inspection
ultrasonic inspection
visual inspection

Weld Defects and Testing

All welds have flaws, or *discontinuities*. The role of inspection is to locate and determine the extent of those flaws. Extensive flaws are called *defects* and may cause a weld to fail. *Nondestructive testing* or inspection of a weld or assembly verifies quality and does not cause damage. The part is usable after the testing is done. *Destructive testing* involves a series of tests by destruction to determine the physical properties of a weld. This type of testing is used to verify welding procedures and qualify welder performance. Only nondestructive inspection methods are discussed in this chapter. These include visual, liquid penetrant, magnetic particle, ultrasonic, and radiographic testing. Chapter 18 further explains criteria used for inspection and welder qualification.

Welding Procedures

Many welding procedures have been created for various welding tasks. Procedures are customized based on the experience of the welder or person developing the procedure and the needs of the application or industry.

Electrode manufacturers have developed procedures for each of their products. An electrode will be used with different material thicknesses, applications, and positions. Since the manufacturer does not know the exact conditions of use, a procedure will yield the minimum and maximum settings for electrode speed and voltage requirements. Other considerations include equipment limitations and experience of the welder applying the procedure. The purpose of using a welding procedure is to obtain a weld quality that meets the specification and to deposit the weld material in a cost-efficient manner.

Test Welds

A procedure must be confirmed by test welds before actual welding begins. A test weld and the required weld should have the same joint design, metal type, and material thickness. The test weld should be inspected for quality on completion. The weld must perform its intended function. Testing can be done by visual inspection, cutting welds apart and using a microscope to view the weld cross section, pulling apart for a tensile test, bending the welded joint in a press, applying a penetrant, or performing one of any number of tests.

Qualified Welding Procedure

Every weldment must meet a standard of quality for the particular weld used. Determining whether a standard of quality has been met can range from a casual look at a weld to see if the parts are joined, to a prescribed inspection for compliance with a specification. High-quality welds may require the use of test welds to establish correct parameters before production begins. Nondestructive testing is performed to verify quality, while destructive testing is performed to verify mechanical values. If the tests are acceptable, the welding procedure is considered a *qualified procedure*, and the product can be welded using that procedure.

If changes are made to the welding values after qualification, testing must be repeated to requalify the new procedure. Anytime you set up a machine, change the electrode, install a gas supply, or modify a parameter, run a test weld on a piece of scrap material. It is far better to throw away a piece of scrap metal than to ruin a production part.

Visual Inspection

Visual inspection is the first and most commonly used type of inspection method. This involves viewing the weld on the front or top surfaces (and penetration side, if visible). In addition, measuring tools, scales, squares, mirrors, a flashlight, and other weld measurement instruments may be used to determine the condition of the weld. See **Figure 17-1**.

The following common defects can be identified by nondestructive visual examination:

Goodheart-Willcox Publisher

Figure 17-1. A variety of tools for measuring, observing, and documenting welds are used during the visual inspection operation.

- Excessive or inadequate crown height.
- Incorrect crown profile.
- Underfill or low weld.
- Undercut.
- Overlap.
- Surface cracks.
- Crater cracks.
- Surface porosity.
- Weld size.
- Weld length.
- Joint mismatch.
- Warpage.
- Variation from dimensional tolerances.
- Inadequate or excessive root side penetration.
- Incorrect root side profile.

The legs of a fillet weld are required to be a specific size. Unless otherwise specified, the leg size should be of equal length, unless otherwise indicated on a print. See **Figure 17-2** and **Figure 17-3**. Fillet welds may also require inspection of the throat dimension, **Figure 17-4**.

Goodheart-Willcox Publisher

Figure 17-2. A fillet weld gauge is used to check the leg size.

Goodheart-Willcox Publisher

Figure 17-3. A fillet weld gauge is used to check if both leg sizes are within tolerance.

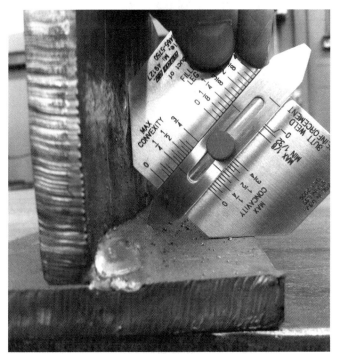

Goodheart-Willcox Publisher

Figure 17-4. Throat size is used to determine concavity or convexity of the weld.

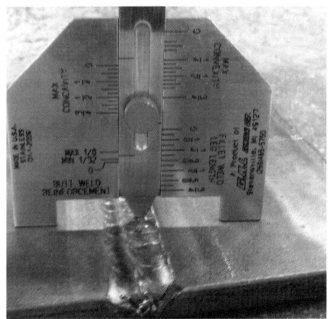

Goodheart-Willcox Publisher

Figure 17-5. This gauge measures the reinforcement on the weld face. This weld face reinforcement is within tolerance.

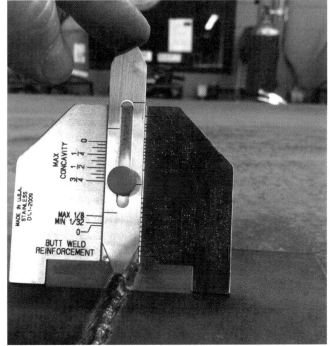

Goodheart-Willcox Publisher

Figure 17-6. The gauge shows that this weld root reinforcement is excessive and unacceptable.

The ripples left by the weld pool should be evenly spaced, without high and low areas or undercut at the weld toe. A special inspection tool can be used to check undercut depth, crown height on a butt weld, and drop-through on the penetration side of the butt weld. See **Figure 17-5** and **Figure 17-6**.

Liquid Penetrant Inspection

Liquid penetrant inspection is a nondestructive test performed on the surface of a weld or the penetration side of a butt weld. Colored liquid or fluorescent dye penetrant is applied to the weld. Some of the penetrant seeps into any cracks or pits, and the penetrant left on the surface is wiped away. Liquid developer is applied, drawing some of the penetrant out of the crevices. The dye permits defects to be seen. See **Figure 17-7** and **Figure 17-8**.

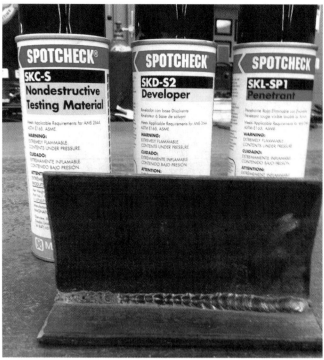

Goodheart-Willcox Publisher

Figure 17-7. A liquid penetrant test was completed on this fillet weld. The red indications show porosity and undercut.

Penetrant inspection does not reveal low welds or undercut. A variation of this inspection technique uses a black light and fluorescent dye. Penetrant processes can be used in any position.

Magnetic Particle Inspection

Magnetic particle inspection is a nondestructive method of detecting cracks, seams, inclusions, segregations, porosity, lack of fusion, and similar discontinuities in magnetic materials. When a magnetic field is established in a *ferromagnetic* material (iron-based with magnetic properties) that contains one or more defects in the path of the magnetic flux, poles are developed at the defects. These poles have a stronger attraction for iron particles than the surrounding material does.

The magnetic field is established by applying electric current to the work, and then a fluid containing iron particles is applied. Any defects are shown by the pattern of the iron particles. See **Figure 17-9** and **Figure 17-10**. Magnetic particle inspection is used mainly for locating defects on the surface of the material. However, this process can locate some defects up to 1/4" below the surface.

Ultrasonic Inspection

Ultrasonic inspection is a nondestructive method of detecting the presence of internal cracks, inclusions, porosity, lack of fusion, and similar discontinuities in

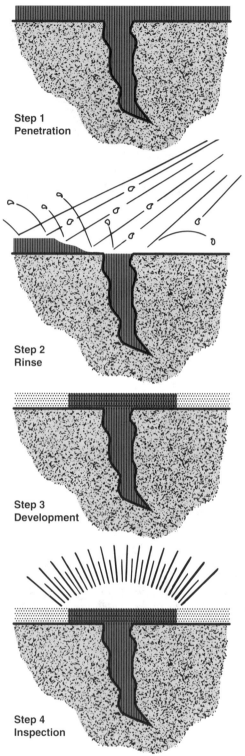

Step 1
Penetration

Step 2
Rinse

Step 3
Development

Step 4
Inspection

Goodheart-Willcox Publisher

Figure 17-8. Liquid penetrant test sequence.

metals. High-frequency sound waves are transmitted through the part. The sound waves return to the sender and appear on a cathode ray tube (CRT). See **Figure 17-11**. Skilled technicians interpret the test results. Recent developments in phased-array ultrasonic

Goodheart-Willcox Publisher

Figure 17-9. A magnetic particle test is being conducted on this butt weld. With the electromagnet activated, the gray test powder will be softly blown away until defect indications remain.

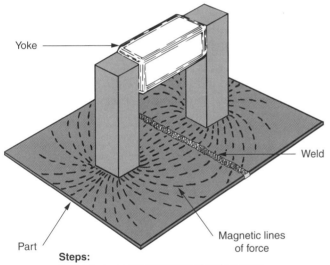

Steps:
1. Apply magnetic field using electric current.
2. Apply magnetic particles while power is on.
3. Blow away excess particles.
4. Inspect.

Goodheart-Willcox Publisher

Figure 17-10. Magnetic particle test sequence.

inspection techniques will result in a safer, more thorough and precise inspection than any other method. Ultrasonic testing has certain advantages, including the following:

- Superb penetration power, permitting testing of thick materials and a variety of welds.
- Sensitivity sufficient to locate very small defects quickly.
- Ability to be done from one surface.

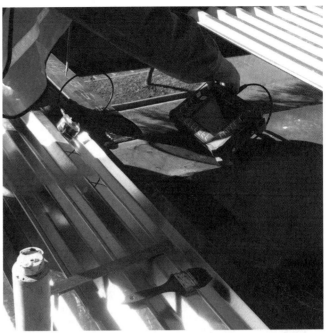

Goodheart-Willcox Publisher

Figure 17-11. This inspector is preparing the weld and equipment for ultrasonic testing. During an ultrasonic test, the transducer is coupled (connected) to the test area by a thin layer of liquid.

Radiographic Inspection

Radiographic inspection is another nondestructive test that shows the presence and nature of discontinuities inside a weld. X-rays and gamma rays penetrate the weld, and flaws are revealed on exposed radiographic film or electronic display, **Figure 17-12**. Although radiographic inspection is expensive compared to other types of tests, the film or printout of the display creates a permanent record of the weld quality.

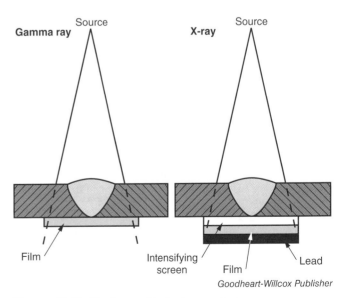

Goodheart-Willcox Publisher

Figure 17-12. Radiographic test operation.

Weld Defects and Corrective Action

Certain defects are common to particular welds. Defects affecting groove, fillet, plug, and spot welds are described next, along with suggestions for correcting each problem. Defects and corrective actions are similar for GMAW and FCAW procedures. More than one corrective action may be needed for any given defect.

Groove Weld Defects

Lack of (or incomplete) penetration. A weld that does not properly penetrate into the weld joint.

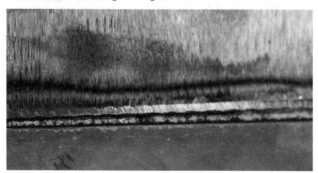

Goodheart-Willcox Publisher

How to correct:
- Open the groove angle.
- Decrease the root face.
- Increase the root opening.
- Increase the amperage.
- Decrease the voltage.
- Decrease the travel speed.
- Change the gun angle.
- Decrease the electrode stickout.
- Keep the arc on the leading edge of the molten pool.

Lack of fusion. A failure of the weld metal to properly join with the base metal or adjoining weld beads.

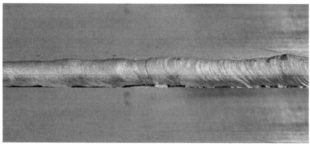

Goodheart-Willcox Publisher

How to correct:
- Clean the weld joint before welding.
- Remove oxides from the previous welds.
- Open the groove angle.
- Decrease the root face.
- Increase the root opening.
- Increase the amperage.
- Decrease the voltage.
- Decrease the travel speed.
- Change the gun angle.
- Decrease the electrode stickout.
- Keep the arc on the leading edge of the molten pool.

Overlap. Weld metal that has flowed over the edge of the joint and improperly fused with the base metal.

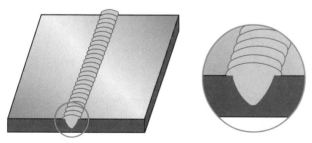

Goodheart-Willcox Publisher

How to correct:
- Clean the edge of the weld joint.
- Remove oxides from the previous welds.
- Reduce the size of the bead.
- Increase the travel speed.

Undercut. Lack of filler material at the toe of the weld.

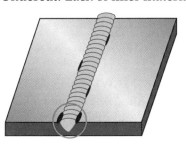

How to correct:
- Decrease the travel speed.
- Increase the dwell time at the edge of the joint on weave beads.
- Decrease the voltage.
- Decrease the amperage.
- Change the gun angle.

Goodheart-Willcox Publisher

Convex crown. A weld that is peaked in the center.

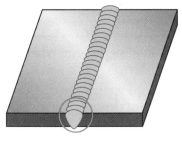

How to correct:
- Change the gun angle.
- Increase the current.
- Decrease the electrode stickout.
- Use a weave bead technique, with a dwell time at the edge of joint.

Goodheart-Willcox Publisher

Craters. Depressions formed at the end of a weld bead due to a lack of filler metal or weld shrinkage.

How to correct:
- Do not stop welding at the end of the joint (use run-off tabs).
- Using the same gun travel angle, move the gun backward a quarter inch, over the completed weld, to backfill the crater before stopping the weld.

Praphan Jampala/Shutterstock.com

Cracks. Fractures caused by cooling stresses in the weld and/or base metal.

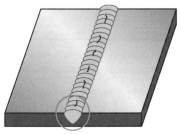

How to correct:
- Use electrode wire with a lower tensile strength or different chemistry.
- Increase joint preheat to slow the weld cooling rate.
- Allow the joint to expand and contract during heating and cooling. Increase the size of the weld.

Goodheart-Willcox Publisher

Porosity. Small internal voids in the weld caused by gas bubbles that did not have enough time to rise to the surface of the molten weld pool before it froze.

Goodheart-Willcox Publisher

How to correct:
- Remove all heavy rust, paint, oil, or scale on the joint before welding.
- Remove any oxide film (GMAW) or slag (FCAW) from the previous passes or layers of the weld.
- Increase the electrode stickout distance for FCAW.
- Check the gas flow.
- Protect the welding area from wind.
- Remove spatter from the interior of the gas nozzle.
- Check the gas hoses for leaks.
- Check the gas supply for contamination.

Linear porosity. Small voids that form in a line along the root of the weld at the center of the joint where penetration is very shallow.

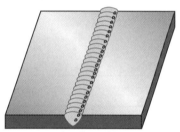

Goodheart-Willcox Publisher

How to correct:
- Make sure the root faces are clean.
- Increase the current.
- Decrease the voltage.
- Decrease the electrode stickout.
- Decrease the travel speed.

Burn-through. A widened gap that forms in the weld joint or a hole that forms where the metal is thin. It is caused by excessive heat in a concentrated area.

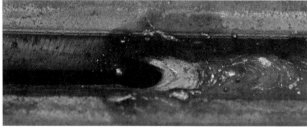

Goodheart-Willcox Publisher

How to correct:
- Decrease the current.
- Increase the voltage.
- Increase the electrode stickout.
- Increase the travel speed.
- Decrease the root opening.

Whiskers. Pieces of electrode wire that extend through the completed weld joint. Occurs where there is a root spacing that the wire slips into.

Goodheart-Willcox Publisher

How to correct:
- Keep the arc on solidified metal and base metal.
- Do not allow the electrode to slip into the root opening.

Excessive penetration. Weld metal that penetrates beyond the bottom of the normal weld root contour.

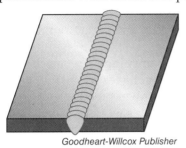
Goodheart-Willcox Publisher

How to correct:
- Decrease the root opening.
- Increase the root face.
- Increase the travel speed.
- Decrease the amperage.
- Increase the voltage.
- Change the gun angle.
- Increase the electrode stickout.

Excessive spatter. Little droplets of filler metal expelled from the weld pool that solidify on the surface of the weld and base metal.

Goodheart-Willcox Publisher

How to correct:
- Decrease the voltage.
- Decrease the drag angle.
- Decrease the travel speed.
- Increase the electrode stickout.
- Decrease the electrode feed speed.
- Use an antispatter spray.
- Do not use CO_2 for GMAW on steel.
- Use the pull technique for GMAW.

Worm tracks. Surface voids usually between 1/4″ to 1″ in length found beneath the slag covering of FCAW-G welds. Shielding gas pockets that are trapped beneath the slag coating as it hardens form these surface voids.

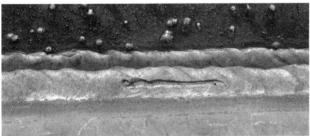
Goodheart-Willcox Publisher

How to correct:
- Decrease the drag angle.
- Decrease the travel speed.
- Increase the electrode stickout.
- Decrease the electrode feed speed.
- Decrease the shielding gas flow rate.
- Use the pull technique for FCAW-G.

Fillet Weld Defects

Lack of penetration. Insufficient weld metal penetration into the joint intersection.

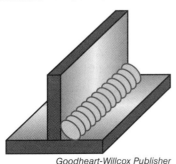
Goodheart-Willcox Publisher

How to correct:
- Decrease the gun travel angle.
- Increase the amperage.
- Decrease the voltage.
- Decrease the size of the weld deposit.
- Use stringer beads.
- Do not make weave beads on root passes.

Lack of fusion. A failure of filler metal to join with the base metal or with previous passes in a multiple-pass weld.

Goodheart-Willcox Publisher

How to correct:
- Remove oxides and scale from the previous weld passes.
- Increase the amperage.
- Decrease the voltage.
- Decrease the travel speed.
- Change the gun's work or travel angle.
- Decrease the electrode stickout.
- Keep the arc on the leading edge of molten pool.

Overlap. A weld face that is larger than, and protrudes over, the weld toe.

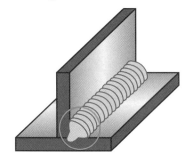 

Goodheart-Willcox Publisher

How to correct:
- Reduce the size of the weld pass.
- Reduce the amperage.
- Change the gun angle.
- Increase the travel speed.

Undercut. A groove formed at the top of the weld bead in horizontal fillet welds due to the base metal melting and not being filled.

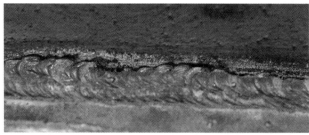

Goodheart-Willcox Publisher

How to correct:
- Make a smaller weld.
- Make a multiple-pass weld.
- Change the gun angle.
- Use a smaller diameter electrode.
- Decrease the amperage.
- Decrease the voltage.

Convexity. A weld that has a high crown.

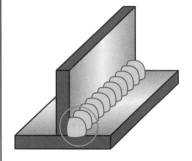 

Goodheart-Willcox Publisher

How to correct:
- Reduce the amperage.
- Decrease the electrode stickout.
- Decrease the voltage.
- Decrease the gun angle.

Craters. A depression formed when weld metal shrinks below the full cross section of the weld.

Goodheart-Willcox Publisher

How to correct:
- Do not stop welding at the end of the joint (use run-off tabs).
- Using the same gun travel angle, move the gun backward a quarter inch, over the completed weld, to backfill the crater before stopping the weld.

Cracks. Fractures in the weld. They occur in fillet welds just as they do in groove welds.

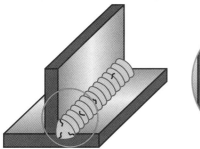

Goodheart-Willcox Publisher

How to correct:
- Use the suggestions given for groove weld cracks.

Burn-through. A hole created when the molten pool melts through the base material.

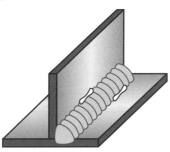

Goodheart-Willcox Publisher

How to correct:
- Decrease the amperage.
- Increase the travel speed.
- Change the gun angle.

Porosity. Small internal voids in the weld caused by gas bubbles that did not have enough time to rise to the surface of the molten weld pool before it froze.

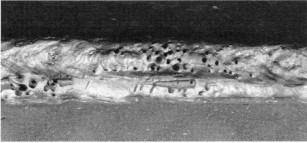

Goodheart-Willcox Publisher

How to correct:
- Use the suggestions provided for groove weld porosity.

Linear porosity. Voids that form along the root of the joint interface.

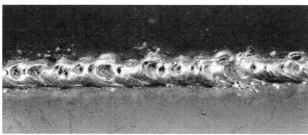
Goodheart-Willcox Publisher

How to correct:
- Use the suggestions provided for groove weld linear porosity.

Worm tracks. Surface voids usually between 1/4″ to 1″ in length found beneath the slag covering of FCAW-G welds. Shielding gas pockets that are trapped beneath the slag coating as it hardens form these surface voids.

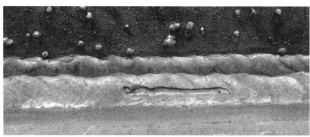
Goodheart-Willcox Publisher

How to correct:
- Use the suggestions provided for groove weld worm tracks.

Plug Weld Defects

Lack of penetration. Failure of the filler metal to reach the proper depth in the bottom plate.

How to correct:
- Increase the amperage.
- Decrease the voltage.
- Decrease the electrode stickout.
- Start the weld in the center of the plug hole, and fill it in a circular pattern.

Excessive penetration. A weld root that extends too far into the bottom sheet of the assembly.

How to correct:
- Decrease the amperage.
- Increase the voltage.
- Increase the electrode stickout.
- Move the gun in a circular pattern and shorten the duration of the welding operation.

Cracks. Fractures that occur at the center of the weld nugget and are caused by rapid cooling of weld metal.

How to correct:
- Use an electrode of a lower tensile strength or different chemistry.
- Increase the preheat to slow the cooling rate.
- Increase the size of the weld.

Porosity. Small internal voids in the weld caused by gas bubbles that did not have enough time to rise to the surface of the molten weld pool before it froze.

How to correct:
- Use the suggestions provided for groove weld porosity.

Overlap. Excessive weld metal that mushrooms out onto the top plate. Caused by too much metal being deposited.

Goodheart-Willcox Publisher

How to correct:
- Shorten the welding time so less metal is deposited into the plug weld hole.

Craters. Depressions that occur in the center of the plug weld when not enough metal is placed in the hole during the weld cycle.

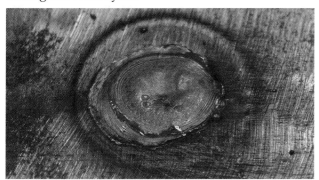

Goodheart-Willcox Publisher

How to correct:
- Lengthen the weld cycle.

Spot Weld Defects

Lack of penetration. Failure of the weld metal to reach the proper depth in the bottom plate.
How to correct:
- Increase the amperage.
- Increase the weld time.
- Decrease the voltage.
- Decrease the electrode stickout.

Excessive penetration. Weld metal that penetrates through the bottom plate.
How to correct:
- Decrease the amperage.
- Decrease the weld time.
- Increase the voltage.
- Increase the stickout.

Porosity. Small internal voids in the weld caused by gas bubbles that did not have enough time to rise to the surface of the molten weld pool before it froze.
How to correct:
- Increase the arc time.
- Increase the amperage.
- Decrease the voltage.
- Decrease the stickout.
- Make sure the metal is clean before mating the assembly.

Cracks. Fractures that occur for the same reasons they do in plug welds.
How to correct:
- Use the suggestions provided for plug weld cracks.

Summary

- All welds have discontinuities (flaws). A flaw becomes a defect when it affects the usability and performance of the product.
- A test weld and the required weld should have the same joint design, metal type, and material thickness.
- Two types of inspection are used for weld testing. Nondestructive testing of a weld or assembly verifies quality without causing damage to the weldment. Destructive testing determines the physical properties of a weld and is used to verify welding procedures and qualify welders.
- Nondestructive inspection methods include visual, liquid penetrant, magnetic particle, ultrasonic, and radiographic inspection. These methods can identify both surface and internal defects.
- Certain defects are common to particular welds. These defects and corrective actions are similar for GMAW and FCAW.
- More than one corrective action may be needed for any given defect.

Review Questions

Answer the following questions using the information provided in this chapter.

1. What is a qualified welding procedure?
2. _____ inspection is done by looking at the weld to determine surface irregularities.
3. _____ inspection uses an electric current and ferromagnetic material to detect discontinuities.
4. _____ inspection uses high-frequency sound waves for locating discontinuities.
5. Weld metal that has flowed over the edge of the joint and improperly fused with the base metal is referred to as _____.
6. How can whiskers be avoided?
7. Molten metal that is expelled from the weld pool is called _____.
8. Pieces of weld wire that extend through the weld joint are called _____.
9. What are three ways to avoid burn-through in a fillet weld?
10. What are three ways to prevent cracking in a plug weld?

Chapter

Welding Codes, Procedures, and Qualifications

Objectives

After studying this chapter, you will be able to:

❑ Explain the relationship between a WPS and a PQR.

❑ Identify the three major codes that govern welding.

❑ Describe the destructive weld tests used to qualify a WPS or a welder performance test.

❑ Explain the procedures used for welding procedure testing and welder qualification.

Technical Terms

American Society of Mechanical Engineers (ASME) code
essential variables
fillet weld break test
guided bend test
hardness test
macroetch test
nonessential variables
prequalified procedures
procedure qualification record (PQR)
side bend test
transverse tension test
welder performance qualification test record
welding procedure specification (WPS)

Introduction to Welding Codes, Procedures, and Qualifications

The qualification or certification of welding personnel starts with the appropriate welding code. A welding *code* is a national standard that holds legal standing. The specific code to be used for production welding is specified in the contract documents for the welded items. The code shows the customer the quality standards that are to be met in producing the welds.

Several welding codes have been produced by various organizations, based on the type of welding to be done. A *welding procedure specification (WPS)* is a document that provides the required welding variable information for a specific weld application to ensure that the weld can be duplicated by properly trained welders. The WPS is established based on information from the welding code and the intended use of the product. The WPS is then tested and qualified using various methods to ensure that the welding procedure used will produce welds of acceptable quality. The result of these tests, along with all variables used to produce the test weld, is documented on a *procedure qualification record (PQR)*.

After the procedure has been tested and qualified, any welder who will be performing the welding must use the WPS to complete a test weld. This test weld then undergoes additional examination procedures. If the test weld passes, a *welder performance qualification test record* is produced. The welding inspector signs and stamps the welding qualification test record to certify that the welder is qualified to use the WPS in production situations.

Specifications govern the quality of the completed weldment. Carefully consider each weld you make or for which you are responsible. The safety of many people is at stake if the weld is not made to the rules of the specification. Shortcuts in quality in order to lower production costs or decrease production time cannot be tolerated. Welders must know the requirements of the specification and make each weld to those requirements. Keep in mind that specifications can change often as new materials, processes, variations, and inspection procedures are introduced.

Buildings, structures, ships, pressure vessels, pipelines, bridges, nuclear reactors, and many other applications require sound welds that are made to a proven procedure by a qualified welder. Several codes are currently in use for these application areas and are available for purchase.

Welding Codes

Three major codes governing welding are the *American Society of Mechanical Engineers (ASME) Boiler Code*, the *American Welding Society (AWS) Structural Welding Code*, and the *American Petroleum Institute (API) Pipeline Welding Code*. Additional welding codes have been established for military use, specific industries, and local governments. The welding or project engineer refers to these codes when necessary.

American Society of Mechanical Engineers

The *American Society of Mechanical Engineers (ASME)* has developed eleven separate code sections pertaining to the design, manufacturing, inspection, and installation of boilers, pressure vessels, and nuclear power plants. A separate code addresses pressure piping applications. Section IX *Welding and Brazing Qualifications* of the *ASME Boiler and Pressure Vessel Code* governs the qualification requirements for welding of boilers and pressure vessels and, along with the B31 codes for pressure piping, requires testing and qualification of all welding procedures and welders. The code does not specifically restrict the base materials, filler materials, or welding processes that can be used; however, it requires weld tests to verify the strength and quality of the resulting weld. Other sections of the ASME Code (Section VIII *Pressure Vessels* or Section IV *Heating Boilers*) may state different or additional specifications to the requirements of Section IX. Those requirements would then take precedent for fabrication within that section.

American Welding Society

The American Welding Society (AWS) has developed welding codes primarily for structural applications. The codes are specified for structural steel, stainless steel, aluminum, sheet metal, bridges, and similar applications.

AWS D1.1 *Structural Welding Code—Steel* applies to the construction of buildings and similar structures using only carbon and low-alloy steels, with some limitations on welding processes. This code provides the required information to design and weld structural applications. The AWS code additionally permits use of specified weld joint designs with various processes, without requiring qualification tests for mechanical values when base material and filler material combinations are specified in the code. These are called ***prequalified procedures***. These procedures have been tested and qualified for use by AWS, and a PQR is not needed when these procedures are used. See **Figure 18-1**. There are no prequalified procedures in the AWS code for GMAW-S.

Any welding procedure that is not prequalified may be developed and tested. The results of those tests are indicated on a PQR. Mechanical tests include

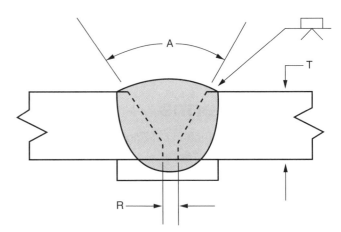

Basic Requirements for Prequalified Joint

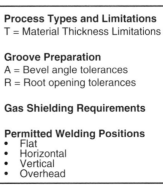

Process Types and Limitations
T = Material Thickness Limitations

Groove Preparation
A = Bevel angle tolerances
R = Root opening tolerances

Gas Shielding Requirements

Permitted Welding Positions
• Flat
• Horizontal
• Vertical
• Overhead

Goodheart-Willcox Publisher

Figure 18-1. AWS prequalified weld joint for V-groove welds.

tests for metal strength, penetration, ductility, and hardness. Other tests may be required as well in order to prove the welding procedure. Changes of variables beyond the code require complete retesting of the procedure and a new WPS and PQR. Production welding is allowed using only a proven WPS and the variables specified in the code.

The variables specified in the welding codes are divided into two categories—essential and nonessential. *Essential variables* are changes that alter the mechanical properties of the completed weldment. These changes require requalification of the WPS and a new PQR. Essential variables include changes to the following:

- Base metal.
- Preheat and postheat treatment.
- Welding process.
- Modes of deposition (GMAW).
- Self-shielded or gas-shielded (FCAW).
- Groove weld with or without backing.
- Welding position.
- Electrode designations.
- Shielding gas types.

Nonessential variables are variables in the welding procedure that have no significant impact on the mechanical properties of the completed weld. These can be changed without requalification of the PQR, but must be noted on the welder performance qualification test record. Nonessential variables include the following:

- Changes to the diameter of the electrode wire.
- Minor changes to voltage and wire speed (amperage) settings.
- Base metal changes within the designated metal group.
- Change in welding direction.

American Petroleum Institute

The American Petroleum Institute (API) has developed welding codes primarily for pipelines and related facilities (API 1104), welding and construction of low-pressure storage tanks (API 620, 650, and 653), and weld inspection and metallurgy (API 577). The API has also established several other standards relating to the petroleum industry.

Destructive Weld Testing

Destructive testing methods are employed to qualify a WPS or evaluate a welder performance test. These methods cannot be used in production welds because the part is damaged or destroyed for

inspection purposes. After the weld passes a visual inspection, the following destructive tests are often used for weld qualifications.

In a *guided bend test*, weld samples or weld coupons are bent in a test machine to a radius specified by the code. This test may be for the weld face and/or weld root. If the material is too thick, a *side bend test* is used. In this test, the coupon weld is centered in the fixture with the side of the weld cross section facing upward. The ram of the test fixture is closed until the legs of the coupon are even. After bending, the coupons are evaluated for cracks on the outer surface of the bent coupon.

Bend tests help determine the ductility and soundness of the weld joint. The guided bend test is the most commonly used destructive test for PQR and welder performance qualification tests. See **Figure 18-2**.

A *fillet weld break test* is performed on a T-joint that has a fillet weld on one side only. A load is applied with a press to the nonwelded side and increased until the weld breaks. This test helps determine joint fusion at the root and toe of the fillet weld. The failed weld is inspected for the presence of any weld discontinuities. See **Figure 18-3**.

A *macroetch test* is often used along with the fillet weld break test. This test typically involves the removal of small samples of the welded joint. When used with the fillet weld break test, the samples are the

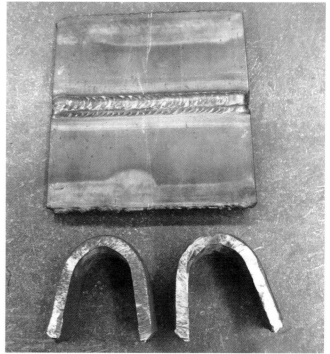

Goodheart-Willcox Publisher

Figure 18-2. A GMAW groove weld test and guided bend test samples.

Figure 18-3. A completed weld that has passed a fillet weld break test.

1″ ends of the test weld. These samples are polished across their cross section and then etched with some type of mild acid mixture, depending on the base material used. The acid etch allows a clear visual appearance of the internal structure of the weld, along with the depth of penetration into the base material. See **Figure 18-4**. This test may also be used to determine depth of penetration for surfacing welds and other joint configurations.

A *transverse tension test* is done to establish the tensile strength and properties of the weld sample. The weld coupon is machined to a narrow section before insertion into the tensile test machine. The weld is then pulled apart and stretched until it fails. This test determines the ductility of the weld, the heat-affected zone, and the weld fusion. The yield limit and tensile strength of the base material and the weld metal are

also determined. This test is usually required as part of the PQR for groove welds.

Hardness tests measure a metal's hardness and resistance to indentation. A small diamond or hardened steel point is pressed into the metal using a specified force. The point is pressed into the weld area at various locations, such as the weld metal, heat-affected metal, and base metal. The hardness value at each point is displayed on the machine. The locations and values are then recorded on the PQR. This test helps determine the extent of the heat-affected zone after the weld is completed and cooled. See **Figure 18-5**.

A nick break test is sometimes required under API Code 1104 for pipe welds. A reduced section of the weld coupon with at least a 3/4″ cross section is nicked or cut slightly so the part will break when twisted and bent. This test allows inspection of discontinuities inside of the weld area.

A Charpy impact test is done to determine the toughness of a weld for low-temperature conditions. This test is always performed in a lab environment with specialized equipment and trained technicians. A small piece of the weld sample is cooled to a very

Figure 18-4. A completed fillet weld test piece before (left) and after (right) the macroetch test.

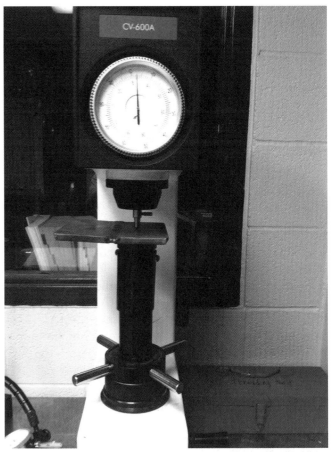

Figure 18-5. A hardness test is being performed on a GMAW groove weld.

low temperature in a liquid nitrogen bath. The notched weld coupon is held in a vise, and a pendulum device with a weighted head is swung to impact the weld. The amount of impact necessary to break the coupon determines the toughness value.

Welder Performance Qualification Test

The procedures described in this chapter can be used for welding procedure testing and welder qualification. The welder performance qualification test record (WPQR) is a document that contains the information from the applicable WPS and PQR used to qualify the welder to that procedure, including the results of the mechanical tests.

Each welding code requires qualification for welders, welding operators, and tackers who will be welding under the code requirements. Qualification tests for GMAW and FCAW welders include the following:

- Plate groove welds and position of the weld.
- Pipe groove welds and position of the weld.
- Plate fillet welds and position of the weld.
- Pipe fillet welds and position of the weld.
- Plate plug and slot welds and position of the weld.
- Tack welds and position of the weld.

Some tests also require qualification for the minimum and maximum thickness of the material to be welded. Welders who qualify on a groove weld test are also qualified for fillet welds with the same thickness and position restriction. See **Figure 18-6**.

Welder Performance Qualification Weld Test Types and Positions

Groove weld test plates for unlimited welding thickness are 1″ thick and qualify for welding 3/16″ thickness and greater. The prepared test plate for unlimited thickness groove welds in the 3G position is shown in **Figure 18-7**. Groove weld test plates for limited thickness are 3/8″ thick and qualify for welding 3/16″ minimum and 3/4″ maximum. The joint design for tubular butt joints for welder qualification is shown in **Figure 18-8**.

Fillet weld test plates are 1/2″ thick and qualify all thicknesses over 3/16″. The fillet weld plate test for welder qualification for unlimited thickness in the 2F position is shown in progress in **Figure 18-9**. The test plate for welder qualification for unlimited thickness fillet welds in the 3F position is being

Welder Qualification—Type and Position Limitations						
Qualification Test		**Type of Weld and Position of Welding Qualified***				
Weld	**Plate or pipe positions**	**Plate**		**Pipe**		
		Groove	**Fillet**	**Groove**	**Fillet**	
Plate-groove	1G	F	F,H	F	F,H	
	2G	F,H	F,H	F,H	F,H	
	3G	F,H,V	F,H,V		F,H	
	4G	F,OH	F,H,OH		F	
	3G & 4G	All	All		F,H	
Plate-fillet	1F		F		F	
	2F		F,H		F,H	
	3F		F,H,V			
	4F		F,H,OH			
	3F & 4F		All			
Pipe-groove	1G	F	F,H	F	F,H	
	2G	F,H	F,H	F,H	F,H	
	5G	F,V,OH	F,V,OH	F,V,OH	F,V,OH	
	6G					
	2G & 5G					
	6GR	All	All	All	All	
	6GR		All		All	
Pipe-fillet	1F		F		F	
	2F		F,H		F,H	
	2F Rolled		F,H		F,H	
	4F		F,H,OH		F,H,OH	
	4F & 5F		All		All	
*Positions of welding: F=flat, H=horizontal, V=vertical, OH=overhead						

Goodheart-Willcox Publisher

Figure 18-6. AWS welder qualification limitations for plate groove and plate fillet welds.

Goodheart-Willcox Publisher

Figure 18-7. Unlimited thickness plate V-groove weld test ready to weld in 3G position.

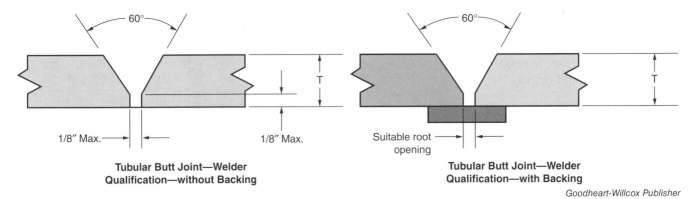

Tubular Butt Joint—Welder
Qualification—without Backing

Tubular Butt Joint—Welder
Qualification—with Backing

Goodheart-Willcox Publisher

Figure 18-8. Joint designs for tubular butt joints for welder qualification.

Goodheart-Willcox Publisher

Figure 18-9. A fillet weld test joint for welder qualification in progress.

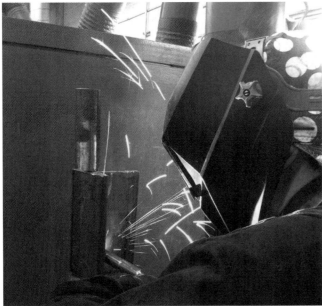

Goodheart-Willcox Publisher

Figure 18-10. Unlimited thickness plate fillet weld test in 3F position.

welded in **Figure 18-10**. Qualification criteria for plate thicknesses below 3/16″ are detailed in AWS specification D1.3 *Structural Welding Code—Sheet Steel*. See **Figure 18-11**.

Welder Performance Qualification Test for Groove Welds

Complete the following steps to practice for an actual welder performance qualification test for GMAW and FCAW on mild steel plate for unlimited thickness:

1. Use the groove weld test WPS shown in **Figure 18-12**. Note: If a radiographic test for quality rather than destructive testing is to be done, guided bend tests are not required, and the backing bar should be 3″ wide instead of the normal 1″ wide steel bar.
2. Remove all slag and wire brush the material.
3. Assemble the plates on the backing bar with a 1/8″ gap, and tack weld each side of the test piece to the backing bar, **Figure 18-13**.
4. Turn the assembly over and make a complete weld on each side of the backing bar. This weld reduces the amount of distortion of the completed assembly and must be done if the bend test specimens are to be oxyacetylene-cut after welding is completed.
5. Place the completed assembly in the desired position for welding.
6. Complete the weld per the WPS, using the proper welding parameters. Make sure the crown dimensions are within specifications. During welding, watch for possible test weld flaws.

Prevent problems by following these suggestions:

- Make a test weld after you have set up the machine to be sure all the settings are correct.
- Use the proper lead or drag angle on the torch.
- Maintain the proper electrode stickout throughout the weld.

Number and Type of Specimens and Range of Thickness Qualified-Welder Qualification											
Plate Test			Number of Specimens					Plate Thickness Qualified		Dihedral Thickness Qualified	
Type of weld	Thickness of test plate (T) as welded	Visual inspection	Bend Test		Side	T-joint break	Macroetch test	Min.	Max.	Min.	Max.
			Face	Root							
Groove	3/8″	Yes	1	1	—	—	—	1/8″	3/4″		
Groove	3/8″<T<1″	Yes	—	—	2	—	—	1/8″	2T		
Groove	1″ or over	Yes	—	—	2	—	—	1/8″	Unlimited		
Fillet Option No. 1	1/2″	Yes	—	—	—	1	1	1/8″	Unlimited	60°	135°
Fillet Option No. 2	3/8″	Yes	—	2	—	—	—	1/8″	Unlimited	60°	135°
Plug	3/8″	Yes	—	—	—	—	2	1/8″	Unlimited		

Figure 18-11. Welder performance qualification test specimen requirements and thickness qualified.

GMAW/FCAW Groove Weld Procedure   GMAW-FCAW 1G, 2G, 3G, 4G

Welding Procedure Specifications: Per AWS D1:1 2010

Welding Process. GMAW root, FCAW Fill & Cap
Position. All
Vertical Progression. Root Pass uphill, Fill & Cap uphill
Joint Type. Butt
Backing: Yes
Backing Material. 1/4″ × 1″ ASTM A36
Root Opening. 1/8″ +/− 1/16″
Groove Angle. 35°–40°
Back Gouge. Only to remove backing material
Base Metal. ASTM A36
Type or Grade. Steel

Thickness. Groove. (in) 1.00
Filler Metals AWS Classification.
GMAW—ER70S-6 FCAW—E71T-1
Shielding Gas. 75% Argon/25% CO_2
Gas Flow Rate. 25–30 cfh
Gas Cup Size. 5/8
Contact Tube to Work Distance. 5/8 to 3/4
Electrical Characteristics. DCEP
Stringer or Weave Bead. Either
Interpass Cleaning. Mechanical or wire brush
Electrode Diameter. GMAW .035 / FCAW .045
Voltage. GMAW 18–22 / FCAW 22–24

Figure 18-12. Welding procedure specification for a GMAW/FCAW combination groove weld test.

- Maintain the welding parameters within the machine's duty cycle limits.
- Use stringer or weave beads as directed by the WPS.
- Remove slag or silicon deposits from each pass with a wire brush. Inspect the completed pass before continuing.
- Take time to do the job right. Know the final crown height dimensions and stay within these limits.

Welding Inspection

After welding is completed, the test weld is visually inspected to ensure adherence to the specification for weld bead profile and to identify visible defects. Study **Figure 18-14**. Cracks are not acceptable in any welded test plate and are cause for rejection. Undercut next

Figure 18-13. Positioning the backing plate before tacking on the back of an unlimited thickness groove weld test.

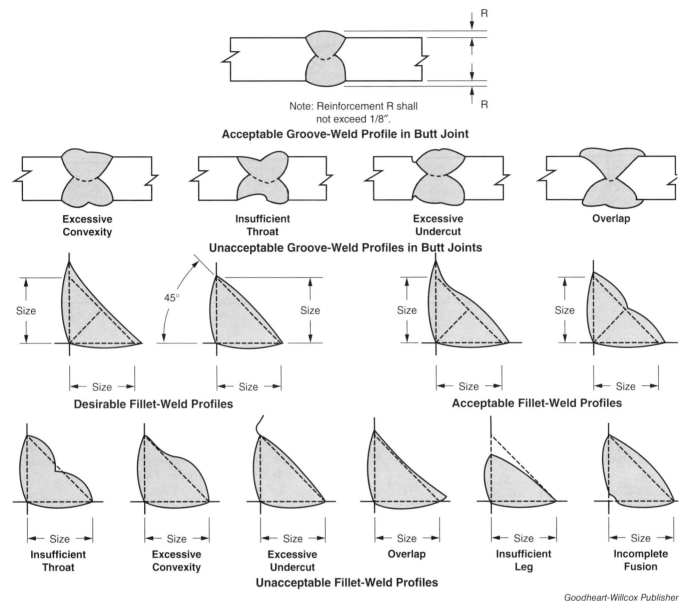

Note: Reinforcement R shall
not exceed 1/8″.
Acceptable Groove-Weld Profile in Butt Joint

Excessive Convexity **Insufficient Throat** **Excessive Undercut** **Overlap**
Unacceptable Groove-Weld Profiles in Butt Joints

Desirable Fillet-Weld Profiles **Acceptable Fillet-Weld Profiles**

Insufficient Throat **Excessive Convexity** **Excessive Undercut** **Overlap** **Insufficient Leg** **Incomplete Fusion**
Unacceptable Fillet-Weld Profiles

Goodheart-Willcox Publisher

Figure 18-14. Acceptable and unacceptable weld profiles for groove and fillet welds.

to the outer edge of the weld bead is limited to 1/32″. Measuring such a small distance is difficult, and the eye can be deceived. Gauges are the best tools for measuring the depth of undercut and the bead profile.

Bend Test

Two 3/8″-thick coupons (specimens) are required from each butt weld test plate for the side bend test. A completed weld test plate with two side bend coupons removed is shown in **Figure 18-15**. If the coupons are cut by a band saw, they are cut to a 3/8″ thickness. If they are cut with an oxyacetylene torch, the coupon must be 5/8″ thick, and 1/8″ must be machined from each side

to achieve a 3/8″ thick coupon. Remove the backing strip with a gouging torch and grind any remaining material from the root side. See **Figure 18-16**. Polish the sides of the coupon, and round the edges. Grind excess crown and remove any backing material from the base of the weld.

The bend test must be made in an AWS-approved test fixture, **Figure 18-17**. The coupon weld is centered in the fixture with the weld cross section facing upward. This is called a *side bend test*. The ram of the test fixture is closed until the legs of the coupon are even. The coupons are evaluated after bending for cracks on the outer surface of the bent coupon, **Figure 18-18**.

Goodheart-Willcox Publisher

Figure 18-15. Test coupons 3/8″ thick have been removed and prepared for side bend testing.

Goodheart-Willcox Publisher

Figure 18-17. Bend test coupons are bent completely around the plunger part of the test fixture.

Goodheart-Willcox Publisher

Figure 18-16. The oxyacetylene-cut and machined coupon with the backing bar removed. This test coupon is ready to bend in an AWS-approved bend test fixture.

Goodheart-Willcox Publisher

Figure 18-18. The outer surface of the test coupon is inspected for cracks. The welded area will be clearly defined after bending.

Inspection and Rejection Criteria

Cracks over 1/8″ long are unacceptable. The length of all cracks under 1/8″ must be added together, and the sum must be equal to or less than 3/8″.

Corner cracks must be evaluated for visible slag or inclusions. Flaws resulting from these areas are limited to 1/8″ maximum to be considered defects. Specimens with corner cracks exceeding 1/4″, with no evidence of slag inclusions or other types of fusion discontinuities, are not used and another sample from the original weldment is prepared and inspected. If there is evidence of slag inclusions or other

discontinuities, the sample is rejected. On completion of nondestructive and destructive tests, acceptance or rejection of each weld test coupon is indicated on the welder performance qualification test record.

If one section of the test fails, the entire test is rejected. A retest can be made under certain conditions. An immediate retest can be made with two tests of the same procedure as used for the failed test. A single retest of the failed weld procedure can be made at a later date after proof of additional instruction.

Fillet Weld Qualification Test

Fillet welding requires a qualified WPS soundness test for procedure qualification. The soundness

Goodheart-Willcox Publisher

Figure 18-19. Unlimited thickness fillet weld test completed. Note the stop and restart in the middle section of the weld.

test defines the maximum single-pass weld size and the minimum multiple-pass bead size for production welding. See **Figure 18-19**.

The weld test for qualification of welders for fillet welds is shown in **Figure 18-20**, using the WPS in **Figure 18-21** and welded in the position required. The following are the steps in a fillet weld qualification test:

The weld is started on one end, stopped in the center, restarted, and then continued to the end. A fillet weld test is marked for macroetch test cuts. This operation should be done on a band saw rather than

using an oxyacetylene cutting torch. Polish and etch for weld size and penetration.

1. A T-joint is set up using two pieces of metal $1/2'' \times 4'' \times 8''$ long.
2. The weld is deposited on only one side, with a stop and restart located in the center one inch of the coupon.
3. After the weld test is completed, it is visually inspected for undercut and weld size.
4. The two end one-inch sections are removed and prepared for the macroetch test.
5. A macroetch test is then performed (refer to **Figure 18-2**). The penetration must extend to the root of the joint but not necessarily beyond.
6. Welder fillet weld tests also require a break test on the center section of the test plate. The test plates are bent with the weld root in tension until the plates are flat or the weld fails, **Figure 18-3**.
7. If the weld does not fracture, the test is satisfactory. If the weld breaks, the break is examined for complete fusion of the root with no porosity larger than $3/32''$ diameter. The sum of the greatest dimensions of all inclusions must not exceed $3/8''$. Porosity dimensions larger than those allowed require a retest.
8. Results of the nondestructive and destructive tests are recorded on the PQR form as either acceptable or rejected, **Figure 18-22**.

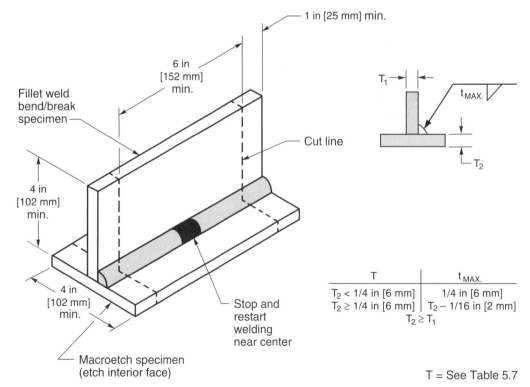

T	$t_{MAX.}$
$T_2 < 1/4$ in [6 mm]	1/4 in [6 mm]
$T_2 \geq 1/4$ in [6 mm]	$T_2 - 1/16$ in [2 mm]
$T_2 \geq T_1$	

T = See Table 5.7

AWS B2.1/B2.1M:2014, Figure 5.8, reproduced with permission of the American Welding Society

Figure 18-20. Unlimited thickness fillet weld plate test.

QW-482 SUGGESTED FORMAT FOR WELDING PROCEDURE SPECIFICATION (WPS)
(See QW-201.1, Section IX, ASME Boiler and Pressure Vessel Code)

Company Name	JM Services - JM Welding	By: Employee # 00123
Welding Procedure Specification No.	JMS - WPS - 0001 Date: 12-17-2013	Supporting PQR No.(s) JMS - PQR - 0001
Revision No. N/A		Date N/A
Welding Process(es)	GMAW	Type(s) Semi-Automatic

(Automatic, Manual, Machine or Semi-Automatic)

JOINTS (QW-402)   Details

Joint Design   Tee Joint
Root Spacing   None
Backing (Yes) _____ (No) XX
Backing Material (Type)   N/A
(refer to both backing and retainers)

☐ Metal   ☐ Nonfusing Metal
☐ Nonmetalic   ☐ Other

FIG A (BUTT JOINT)   FIG B (CORNER JOINT)   FIG C (EDGE JOINT)

Sketches, Production Drawings, Weld Symbols or Written Description should show the general arrangement of the parts to be welded. Where applicable, the root spacing and the details of weld groove may be specified.

[At the option of the manufacturer, sketches may be attached to illustrate joint design, weld layers and bead sequence (e.g., for notch toughness procedures, for multiple process procedures etc.)]

FIG D (LAP JOINT)   FIG E (TEE JOINT)

☒ Other (Describe)   1/2" x 4" x 8" 2 pcs.

☐ Figure A   ☐ Figure B   ☐ Figure C   ☐ Figure D   ☒ Figure E

*BASE METALS (QW-403)

P-No. ____P-1____ Group No. _____ to P-No. ____P-1____ Group No. _____

OR

Specification type/grade or UNS number _____
to Specification type/grade or UNS number _____

OR

Chem. Analysis and Mech. Prop.   N/A
to Chem. Analysis and Mech. Prop.   N/A

Thickness Range:
    Base Metal: Groove   N/A   Fillet   1/4" to 1"
Other   N/A

Maximum Pass Thickness ≤ 1/2 inch (13 mm)   (Yes) ☒   (No) ☐

*FILLER METALS (QW-404)

	1	2	3	4
Spec. No. (SFA)	ESAB MIG-6			
AWS No. (Class)	ER70S-6			
F-No.	6			
A-No.				
Size of Filler Metals	.035			
Filler Metal Product Form	Bare wire			
Supplemental Filler Metal	N/A			
Weld Metal Thickness Range:				
Groove	N/A			
Fillet	1/4"–3/8"			
Electrode-Flux (Class)	N/A			
Flux Type	N/A			
Flux Trade Name	N/A			
Consumable Insert	None			
Other	N/A			

(Continued)

*Each base metal-filler metal combination should be recorded individually.

Figure 18-21. This WPS contains the required information for a welder performance qualification test for fillet welds.

QW-482 (Back)

WPS No.	JWS - WPS - 0001	**Rev** _____

POSITIONS (QW-405)
Position(s) of Groove _____ N/A _____
Welding Progression: Up ___ N/A ___ Down ___ N/A ___
Position(s) of Fillet _____ Horizontal - 2F _____
Other _____

PREHEAT (QW-406)
Preheat Temperature, Minimum _____ 70F _____
Interpass Temperature Maximum _____ 320F _____
Preheat Maintenance _____ N/A _____
Other ___ N/A ___
(Continuous or special heating where applicable should be recorded)

POST WELD HEAT TREATMENT (QW-407)
Temperature Range _____ None _____
Time Range _____ N/A _____
Other _____ N/A _____

GAS (QW-408)

		Percent Composition	
	Gas(es)	(Mixture)	Flow Rate
Shielding	75% AR	25% CO2	20 - 30
Trailing	None		
Backing	None		
Other	None		

ELECTRICAL CHARACTERISTICS (QW-409)

_____ DCEP _____
_____
_____
_____

Weld Pass(es)	Process	Classification	Diameter.	Type and Polarity	Amperage Range	Voltage Range	Travel Speed Range	Other (e.g., Remarks, Comments, Hot Wire Addition, Technique, Torch Angle, Etc.)
1	GMAW	ER70S-6	.035	DCEP	100 - 140	18 - 19.5	6 -8 ipm	Forehand
2	GMAW	ER70S-6	.035	DCEP	100 - 140	18.- 19.5	6 - 8 ipm	Forehand
3	GMAW	ER70S-6	.035	DCEP	100 - 140	18.- 19.5	6 - 8 ipm	Forehand

(Amperage and voltage range should be recorded for each electrode size, position, and thickness, etc)

Pulsing Current _____ N/A _____   Heat Input (max) _____

Tungsten Electrode Size and Type _____ N/A _____
(Pure Tungsten, 2% Thoriated, etc.)

Mode of Metal Transfer for GMAW _____ Short Circuiting _____
(Spray arc, short circuiting arc, etc.)

Electrode Wire feed speed range _____ 230 - 250 ipm _____

Other _____

TECHNIQUE (QW-410)
String or Weave Bead _____ Stringer _____
Orifice or Gas Cup Size _____ 5/8" _____
Initial and Interpass Cleaning (Brushing, Grinding, etc.) _____ Wire Brush _____
Method of Back Gouging _____ N/A _____
Oscillation _____ yes _____
Contact Tube to Work Distance _____ 1/4"- 3/8" _____
Multiple or Single Pass (per side) _____ Multiple _____
Multiple or Single Electrodes _____ Single _____
Electrode Spacing _____ N/A _____
Peening _____ N/A _____
Other _____ None _____
_____ N/A _____

Figure 18-21. *(Continued)*

QW-483 SUGGESTED FORMAT FOR PROCEDURE QUALIFICATION RECORDS (PQR)
(See QW-200.2, Section IX, ASME Boiler and Pressure Vessel Code)
Record Actual Conditions Used to Weld Test Coupon

Company Name	JM Services - JM Welding
Procedure Qualification Record No.	JMS - PQR -0001 Date 12-17-2013
WPS No.	JMS - WPS - 0001
Welding Process(es)	GMAW
Types (Manual, Automatic, Semiautomatic)	SemiAutomatic

JOINTS (QW-402)

Groove Design of Test Coupon (sketch, figure or reference)

(For combination qualifications, the deposited weld metal thickness shall be recorded for each filler metal and process used.)

BASE METALS (QW-403)

Material Specification	
Type or Grade or UNS Number	
P No. 1 Group No.	to P No. 1 Group No
Thickness of Test Coupon	1/2"
Diameter of Test Coupon	N/A
Maximum Pass Thickness	3/8"
Other	N/A

FILLER METALS (QW-404)

	1	2	3
Layer (combination welds)	1	2	3
SFA Specification			
AWS Classification	ER70S-6	ER70S-6	ER70S-6
Filler Metal F No.	F6	F6	F6
Weld Metal Analysis A No.			
Size of Filler Metal	.035	.035	.035
Filer Metal Product Form	Bare Wire	Bare Wire	Bare Wire
Supplemental Filler Metal	N/A		
Electrode Flux Classification	N/A		
Flux Type	N/A		
Flux Trade Name	N/A		
Weld Metal Thickness	3/8"	3/8"	3/8"
Other			

POSITION (QW-405)

Position of Groove	N/A
Weld Progression (Uphill, Downhill)	N/A
Other	N/A

PREHEAT (QW-406)

Preheat Temperature	70°F
Interpass Temperature	
Other	N/A

POST WELD HEAT TREATMENT (QW-407)

Temperature	N/A
Time	N/A
Other	N/A

GAS (QW-408)

	Gas (es)	Percent Composition (Mixture)	Flow Rate
Shielding	Ar/CO$_2$	75/25	20–30
Trailing	N/A		
Backing	N/A		
Other	N/A		

ELECTRICAL CHARACTERISTICS (QW-409)

Current	DC
Polarity	EP
Amps	Volts
Tungsten Electrode Size	N/A
Transfer Mode for GMAW (FCAW)	Short Circuit
Other	N/A

TECHNIQUE (QW-410)

Travel Speed	6"– 8" per min.
String or Weave Bead	Stringer
Oscillation	Yes
Multipass or Single Pass (per side)	Multipass - 3 beads
Single or Multiple Electrodes	Single
Other	N/A

Figure 18-22. After testing is done, the PQR is completed and indicates either acceptance or rejection. The second page of the PQR form contains sections for entering results of tests such as tensile, guided bend, toughness, and fillet weld tests.

Welder Qualification and Certification

Upon successful completion of any welding test, a welder performance qualification test record is completed. The inspector, certifying that the statements on the record are correct, signs a form, **Figure 18-23**. The terms *qualified welder* and *certified welder* describe the person who has completed an acceptable weld test.

The AWS code considers a welder to be qualified indefinitely, with the following exceptions:

AWS D1.1/D1.1M:2015   ANNEX M

Sample Welder Qualification Form (Single-Process)
WELDER, WELDING OPERATOR, OR TACK WELDER
PERFORMANCE QUALIFICATION TEST RECORD

Name		OPTIONAL PHOTO ID	Test Date		Rev.
ID Number			Record No.		
Stamp No.			Std. Test No.		
Company			WPS No.		
Division			Qualified To		

BASE METALS	Specification	Type or Grade	AWS Group No.	Size (NPS)	Schedule	Thickness	Diameter
Base Material							
Welded To							

VARIABLES	Actual Values		RANGE QUALIFIED	
Type of Weld Joint				
Base Metal				

	Groove	Fillet	Groove	Fillet
Plate Thickness				
Pipe/Tube Thickness				
Pipe Diameter				

Welding Process	
Type *(Manual, Semiautomatic, Mechanized, Automatic)*	
Backing	
Filler Metal (AWS Spec.)	
AWS Classification	
F-Number	
Position	
Groove – Plate & Pipe ≥ 24 in	
Groove – Pipe < 24 in	
Fillet – Plate & Pipe ≥ 24 in	
Fillet – Pipe < 24 in	
Progression	
GMAW Transfer Mode	
Single or Multiple Electrodes	
Gas/Flux Type	

TEST RESULTS

Type of Test	Acceptance Criteria	Results	Remarks

CERTIFICATION

Test Conducted by	
Laboratory	
Test Number	
File Number	

We, the undersigned, certify that the statements in this record are correct and that the test welds were prepared, welded, and tested in accordance with the requirements of Clause 4 of AWS D1.1/D1.1M (_____) *Structural Welding Code—Steel*.
                                                                     (year)

Manufacturer or Contractor _____   Authorized by _____

                                                          Date _____

AWS D1.1/D1.1M:2015, Annex N reproduced with the permission of the American Welding Society, Miami, FL

Figure 18-23. Welder performance qualification test results are recorded on a test record. The test data used for the qualification test are indicated on the form. The form indicates whether the welder passed or failed the test.

- The welder is not engaged in a process of welding for which he or she is qualified for a period exceeding six months. (A requalification test may be made on 3/8″ material.)
- There is a specific reason to question the welder's ability.

Tackers, Welding Operators, and Other Qualification Tests

The tests for tackers and welding operators, and other types of welding tests specified in the code, are not presented in this text. For information regarding these qualification tests, contact the American Welding Society regarding *AWS D1.1 Structural Welding Code—Steel*.

Kimtaro/Shutterstock.com

Weld coupons are bent in a test machine to evaluate the ductility and soundness of the weld joint.

Summary

- The qualification or certification of welding personnel starts with the appropriate welding code. A welding code is a national standard that holds legal standing.
- A welding procedure specification (WPS) is established based on information from the code. This WPS is then tested and qualified using various methods, and this information is documented on a procedure qualification record (PQR).
- Three major codes governing welding are the *American Society of Mechanical Engineers (ASME) Boiler Code*, the *American Welding Society (AWS) Structural Welding Code*, and the *American Petroleum Institute (API) Pipeline Welding Code*.
- Variables specified in the welding codes are either essential variables (changes that alter the mechanical properties of the completed weldment) or nonessential variables (changes that have no significant impact on the mechanical properties of the completed weld).
- Changes to essential variables require qualification of a new welding procedure specification. Changes to nonessential variables do not require a new welding procedure specification.
- Welding codes require qualification for welders, welding operators, and tackers welding under the code requirements. The welder completes a weld following a WPS. That weld is tested and inspected to the code standards. If the weld passes, the inspector certifies a welder performance qualification test record.
- Destructive testing methods are used when qualifying a WPS or a welder performance test. These methods cannot be used in production welds because the part is damaged or destroyed during the testing procedure.

Review Questions

Answer the following questions using the information provided in this chapter.

1. What is a *welding procedure specification*?
2. What three codes govern welding operations?
3. What is a *prequalified procedure*?
4. Explain the difference between essential and nonessential weld variables.
5. Requalification of the WPS and a new PQR are required when _____ variables are changed.
6. Destructive testing methods are used to qualify a WPS or a(n) _____ test.
7. List seven destructive test methods.
8. Groove weld test plates for unlimited welding thickness are _____ thick and qualify for welding _____ thickness and greater.
9. What is the purpose of a fillet weld soundness test?
10. Under what two conditions can a welder performance qualification retest be performed?

Chapter 19

Welding Employment and Careers

Objectives

After studying this chapter, you will be able to:

- Describe the opportunities for employment in the welding industry.
- Explain the steps required to apply for a job.
- Explain the soft skills required for employment.

Technical Terms

Certified Welding Inspector (CWI)
résumé
soft skills

Preparing for a Welding Career

Training and welding certification is only the first step in your career as a welder. Additional cutting and welding process training is recommended. Your ability to read and understand welding prints as well as fabricate parts will continue to improve with practice.

Innovations and improvements in welding equipment, particularly those for the GMAW and FCAW processes, are continuing at a rapid pace. With these rapid changes in technology it is important to stay current with the industry. Continuing education can be achieved in several ways. The Internet web pages of all major welding equipment manufacturers contain up-to-date information on their latest products. Welding and fabrication magazines detail new innovations in the industry. Continue to upgrade your welding skills on new equipment by enrolling in classes at your local community college or training center.

It is recommended that you join the American Welding Society as a student or early in your career. Visit the AWS website for details about the benefits offered by membership.

Careers

As you enter employment, your skill and interest are your only limitations. Becoming a welder is more than just a job; it is a lifelong career opportunity. Employment in the welding field includes the following:

- **Welder's helper.** Many entry-level welders are teamed up with more experienced welders as they learn the particular job requirements.
- **Journeyman welder.** After the employee understands the expectations of the job and can perform welding tasks with little or no supervision, he or she attains journeyman status.
- **Master welder or lead welder.** This person has experience and skills with many welding processes, equipment, troubleshooting, and manufacturing.
- **Welding supervisor.** The welding supervisor oversees the operation of other welders to ensure that quality and production levels are maintained.
- **Shop foreman.** A shop foreman oversees all aspects of the production facility, often leading several departments besides the welding area.
- **Independent contractor** or **owner.** A contract welder or shop owner provides mobile or shop

services for the customer and is responsible for all aspects of the business. These people may work as individual welders or have several employees. They must have both mastery of the welding trade and diverse business skills, such as the ability to accurately estimate costs, draft job proposals, and manage personnel.
- **Welding inspector.** A welding inspector is responsible for quality assurance by inspecting the welds, writing and qualifying procedures, and qualifying welders. See **Figure 19-1**. A *Certified Welding Inspector (CWI)* has industry experience and has completed training and examinations provided by the American Welding Society.
- **Welding sales.** Individuals with welding knowledge and experience sell equipment and supplies from local welding supply stores or industry manufacturers.

Other careers in the welding field require additional education and knowledge. A deeper understanding of weld processes, math, physics, and metallurgy can lead to the following careers:

- **Welding professor.** These instructors are responsible for the training and advancement of future welders. They remain current with updated information, teach welding theory, and demonstrate proper procedures.
- **Welding engineer.** Some engineers are responsible for the design and application of parts, determining welding processes and procedures for specific applications. Other engineers are involved with research of innovative welding processes, equipment, and future applications.
- **Welding consultant.** A consultant has extensive welding experience and advanced training. Consultants are often hired by companies to help solve process problems or provide specialized training to the employees.

All of these careers begin with a basic understanding of the welding process and application of skills. Additional training, experience, and involvement lead to further opportunities in the welding industry.

Applying for a Job

Once you have acquired the basic welding skills presented in this text, along with other welding and cutting process skills, you may be ready to begin work in the welding industry.

Lisa F. Young/Shutterstock.com

Figure 19-1. A welding inspector checking factory welding equipment.

Many large companies now use online applications, which can be found on company websites or job-search websites. These applications can be completed at home or at an employment agency. It is important to honestly complete all required parts of the application. Be certain your spelling is correct. A neat, thought out application will be noticed by the employer.

When filling out an online application, include key terms for which the employer may search. This will help you stand out from the many other applications the employer will receive. Keywords include job titles, job-specific skills, and soft skills (discussed later in this chapter). Save the application in the file format requested. If a preferred format is not given, it is best to save the application in DOC or PDF file format. This will enable the employer to find specific search terms in your document. Be sure to complete all the fields of the application. Many job-search sites have sample forms on which you can practice before attempting a real application.

A smaller company may require a handwritten application. In this case, bring a copy of your résumé to attach to the application. Fill out the application completely, printing your answers neatly and legibly.

A list of references is required by employers. A reference is a person who can provide and verify information about you to an employer. A reference can be an instructor, school official, previous employer, or any other adult outside your family who knows you well. Obtain at least three references. Contact those individuals prior to listing them as references to get their permission and ensure that their contact information is current and accurate. The references will be contacted if the company is interested in employing you.

If possible, submit a résumé along with your job application. A *résumé* is a brief outline of your education, work and military experience, membership in trade organizations, and other qualifications for work. A well-written résumé can help you get an interview. The résumé should be concise, accurate, and error-free. See **Figure 19-2**.

Interview

After the company has reviewed your application or résumé, they may call you for an interview and weld test. Before the interview, research the company or organization to become familiar with their product line. This will help you ask questions about job requirements and show that you have an interest in their company. Think about how your skills and training have prepared you to meet the requirements of the job so you are prepared to answer questions.

Be sure to arrive at the interview ahead of the requested time. Wear clean clothes that you can weld in, as if you are ready to begin work immediately. Bring your personal welding equipment (helmet, gloves, safety glasses, hand tools, etc.) and be prepared to weld. During the interview, answer all questions as honestly as possible, with concise but detailed answers.

After the interview, you may be required to perform a weld test. Make certain that you clearly understand what is expected. Always ask for some scrap metal to get used to the machine and run some practice beads. Once you feel comfortable with the machine settings, take your time and make the best weld you can. If you make a small mistake, let the inspector know what you did and how you could fix it. They know that you are probably nervous and trying too hard. Remember that you have completed similar weld tests during your training and you know what to do.

Once you pass the weld test and are offered the position, the salary negotiation process begins. This should be the first time you ask about wages! Be prepared to start at a slightly lower wage scale during a probationary period. You need to prove you can do the job and fit in with the other employees. After a period of time, during which you learn how the company works and continue to improve your skills, you will be rewarded with a higher income.

Employment Skills

Technical skills are certainly important for your welding career; however, today's employers are looking for skills beyond welding capabilities. Soft skills and work ethics are just as important for career advancement. A strong work ethic includes the traits of initiative, personal integrity, commitment to a quality product, attention to details, and taking responsibility for all aspects of your work.

Soft skills include interpersonal skills, such as communication skills and ability to be a team member, and self-management skills. Soft skills are desired to the degree that some companies will hire a person with good soft skills and train them for the technical skills. The following points describe some important soft skills sought by employers:

- **Ability to work well with others.** Many welders are teamed up to manufacture products. See **Figure 19-3**. A good attitude increases production and makes for an enjoyable work experience. As a new employee, you have a lot to learn about how the company works and the

Michael J. Garcia

134 Lincoln Street   (915) 555-1234
Wilton, TX 93232   mjgarcia22@e-mail. com

Career Objective
To obtain an entry-level welding technician position in the fabrication industry.

Professional Experience
Heavy Metal Welding, Holloton, TX   August 2015–present
Welder's Helper
- Perform general shop labor, including material cutting and preparation with oxyfuel and plasma systems.
- Cut parts to required length on horizontal bandsaw.
- Tack weld ladder systems with GMAW.
- Help welders with material prep and grinding.

Simpson Supply Co., Wilton, TX   May 2014–August 2015
Parts Clerk
- Worked with customers at parts counter, checked inventory system, and obtained parts.
- Conducted daily and monthly inventory checks.
- General stocking and cleaning throughout store.
- Delivered and picked up parts and equipment.

Education
Associate Degree in Welding Technology   May 2016
Gardendale Community College
- GPA: 3.22/4.0
- Coursework included welding layout and fabrication, SMAW, GMAW, FCAW, and GTAW processes, oxyfuel and plasma cutting
- Obtained level one and level two welding technology certificates: Entry Level Welder, and Advanced Welder.
- Passed AWS welder performance qualification test for GMAW/FCAW and SMAW.
- Member AWS student chapter at GCC.
- Participated in SkillsUSA welding completion in high school.

Community Service
Habitat for Humanity, volunteer, summers of 2014, 2015, and 2016
Wilton Food Bank, volunteer, 2014–present

References
Available upon request.

Goodheart-Willcox Publisher

Figure 19-2. An example of a welder's résumé.

Praphan Jampala/Shutterstock.com

Figure 19-3. Welders often work in teams to manufacture products.

products that are welded. This is the time to watch, listen, and learn.

- **Timeliness and time management.** Arriving for work on time and every day, keeping break times as allowed, and remaining productive throughout the workday will be noticed and rewarded. *Always* arrive at least 10 to 15 minutes before you are to begin work. See **Figure 19-4**. Managing work priorities and meeting deadlines are also important. An employee with this work ethic will be seen as responsible when additional duties arise.

Figure 19-4. Always arrive at least 10 to 15 minutes before you are to begin work. If you arrive at the time work is to begin, or even a few minutes late, your tardiness will be noticed by the management.

- **Verbal communication skills.** The ability to listen and speak effectively, as well as understand technical directions, is important for a welder. Manufacturing facilities are often loud, busy, fast-paced environments. A misunderstanding of directions can result in injury or costly mistakes. Verbal communication is multidirectional between the welder, supervisor, and customer.

Figure 19-5. Welders must be able to read and understand documents and prints.

- **Written communication skills.** The ability to read and write is also important for the welder. Understanding prints and documents for welded parts, reading and following welding procedures, and correctly filling out shop reports are essential in the industry. See **Figure 19-5.** In addition, time or job cards, reports, or records must be completed accurately.

Continually develop your soft skills, just as you continue to practice and improve your welding skills. The welding industry will always need qualified individuals that are willing to work hard, learn new ideas, and advance the industry.

Summary

- Many careers are available in the welding industry. Becoming a welder is a lifelong career opportunity.
- Many large companies now use online applications, while smaller companies may require a handwritten application. Be sure that all required sections of the application are accurately completed.
- A concise, accurate, and error-free résumé can help you get an interview.
- After an interview, a weld test may be required. Be sure you clearly understand what is expected and run some practice beads before taking the test.
- Soft skills and strong work ethics are as important as technical skills for career advancement.
- Soft skills include the ability to work well with others, timeliness and time management skills, and communication skills (verbal and written).

Review Questions

Answer the following questions using the information provided in this chapter.

1. A(n) _____ welder understands the expectations of the job and can perform welding tasks with little or no supervision.
2. What are the responsibilities of a welding supervisor?
3. Someone with industry experience who is interested in quality assurance can become a(n) _____ after completing AWS training and exams.
4. A(n) _____ is a person unrelated to you who can provide information about you to an employer.
5. List two reasons why you should research a company before interviewing with them.
6. What should you do if you make a small mistake during a weld test for a potential employer?
7. When should you ask about wages during the employment interview?
8. Soft skills include interpersonal skills and self-_____ skills.
9. List four important soft skills that employers are seeking.
10. Being on time for work means arriving at least _____ minutes before work is to begin.

Reference Section

The Reference Section contains 27 charts of useful welding-related information. The charts are listed below by title and page number.

Welding Safety Checklist

Hazard	Factors to Consider	Precaution Summary
Electric shock can kill	» Wetness » Welder in or on workpiece » Confined space » Electrode holder and cable insulation	» Insulate welder from workpiece and ground using dry insulation. Rubber mat or dry wood. » Wear dry, hole-free gloves. (Change as necessary to keep dry.) » Do not touch electrically "hot" parts or electrode with bare skin or wet clothing. » If wet area and welder cannot be insulated from workpiece with dry insulation, use a semiautomatic, constant voltage welder or stick welder with voltage reducing device. » Keep electrode holder and cable insulation in good condition. Do not use if insulation is damaged or missing.
Fumes and gases can be dangerous	» Confined area » Positioning of welder's head » Lack of general ventilation » Electrode types, i.e., manganese, chromium, etc. See SDS » Base metal coatings, galvanize, paint	» Use ventilation or exhaust to keep air breathing zone clear, comfortable. » Use helmet and positioning of head to minimize fume in breathing zone. » Read warnings on electrode container and Safety Data Sheet (SDS) for electrode. » Provide additional ventilation/exhaust where necessary to maintain exposures below applicable limits. » Use special care when welding in a confined area. » Do not weld unless ventilation is adequate.
Welding sparks can cause fire of explosion	» Containers that have held combustibles » Flammable materials	» Do not weld on containers that have held combustible materials (unless strict AWS F4.1 procedures are followed). Check before welding. » Remove flammable materials from welding area or shield from sparks, heat. » Keep a fire watch in area during and after welding. » Keep a fire extinguisher in the welding area. » Wear fire retardant clothing and hat. Use earplugs when welding overhead.
Arc rays can burn eyes and skin	» Process: gas-shielded arc most severe	» Select a filter lens that is comfortable for you while welding. » Always use helmet when welding. » Provide nonflammable shielding to protect others. » Wear clothing that protects skin while welding.
Confined space	» Metal enclosure » Wetness » Restricted entry » Heavier-than-air gas » Welder inside or on workpiece	» Carefully evaluate adequacy of ventilation especially where gas may displace breathing air. » If basic electric shock precautions cannot be followed to insulate welder from work and electrode, use semiautomatic, constant voltage equipment with cold electrode or stick welder with voltage reducing device. » Provide welder helper and method of welder retrieval from outside enclosure.
General work area hazards	» Cluttered area	» Keep cables, materials, tools neatly organized.
	» Indirect work (welding ground) connection	» Connect work cable as close as possible to area where welding is being performed. Do not allow alternate circuits through scaffold cables, hoist chains, ground leads.
	» Electrical equipment	» Use only double insulated or properly grounded equipment. » Always disconnect power to equipment before servicing.
	» Engine-driven equipment	» Use in only open, well ventilated areas. » Keep enclosure complete and guards in place. » Refuel with engine off. » If using auxiliary power, OSHA may require GFI protection or assured grounding program (or isolated windings if less than 5kW).
	» Gas cylinders	» Never touch cylinder with the electrode. » Never lift a machine with cylinder attached. » Keep cylinder upright and chained to support.

Adapted from Lincoln Electric Company

Chart R-1. Welding Safety Checklist.

Guide for Shade Numbers				
Operation	**Electrode Size 1/32 in. (mm)**	**Arc Current (A)**	**Minimum Protective Shade**	**Suggested[1] Shade No. (Comfort)**
Shielded metal arc welding	Less than 3 (2.5) 3–5 (2.5–4) 5–8 (4–6.4) More than 8 (6.4)	Less than 60 60–160 160–250 250–550	7 8 10 11	— 10 12 14
Gas metal arc welding and flux cored arc welding		Less than 60 60–160 160–250 250–500	7 10 10 10	— 11 12 14
Gas tungsten arc welding		Less than 50 50–150 150–500	8 8 10	10 12 14
Air carbon arc cutting	(Light) (Heavy)	Less than 500 500–1000	10 11	12 14
Plasma arc welding		Less than 20 20–100 100–400 400–800	6 8 10 11	6 to 8 10 12 14
Plasma arc cutting	(Light)(2) (Medium)(2) (Heavy)(2)	Less than 300 300–400 400–800	8 9 10	9 12 14
Torch brazing		—	—	3 or 4
Torch soldering		—	—	2
Carbon arc welding		—	—	14

	Plate thickness			
	in.	**mm**		
Gas welding Light Medium Heavy	Under 1/8 1/8 to 1/2 Over 1/2	Under 3.2 3.2 to 12.7 Over 12.7		4 or 5 5 or 6 6 or 8
Oxygen cutting Light Medium Heavy	Under 1 1 to 6 Over 6	Under 25 25 to 150 Over 150		3 or 4 4 or 5 5 or 6

(1) As a rule of thumb, start with a shade that is too dark to see the weld zone. Then go to a lighter shade that gives sufficient view of the weld zone without going below the minimum. In oxyfuel gas welding or cutting where the torch produces a high yellow light, it is desirable to use a filter lens that absorbs the yellow or sodium line in the visible light of the (spectrum) operation.

(2) These values apply where the actual arc is clearly seen. Experience has shown that lighter filters may be used when the arc is hidden by the workpiece.

Goodheart-Willcox Publisher

Chart R-2. Guide for shade numbers.

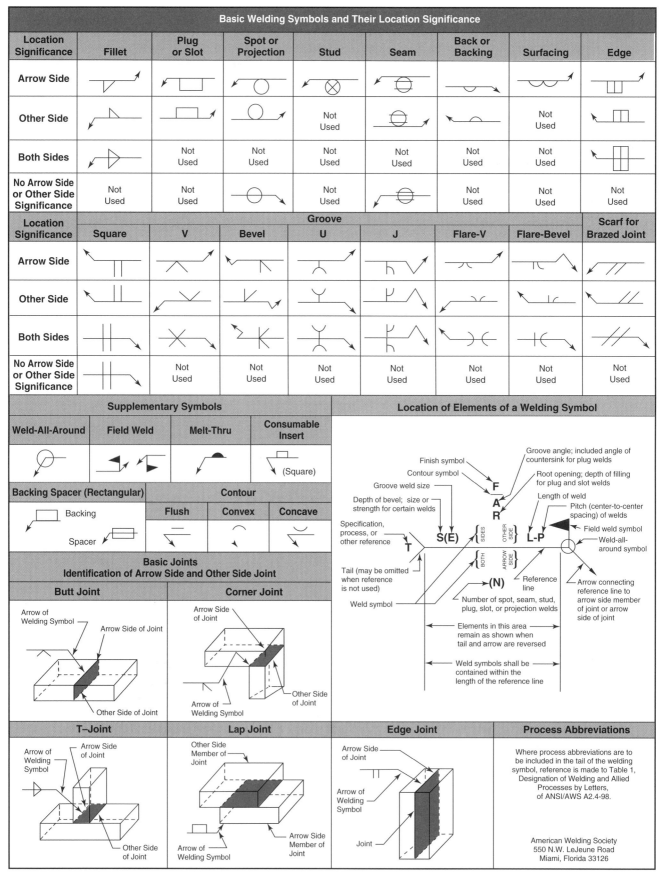

Chart R-3. Basic welding symbols and their location significance.

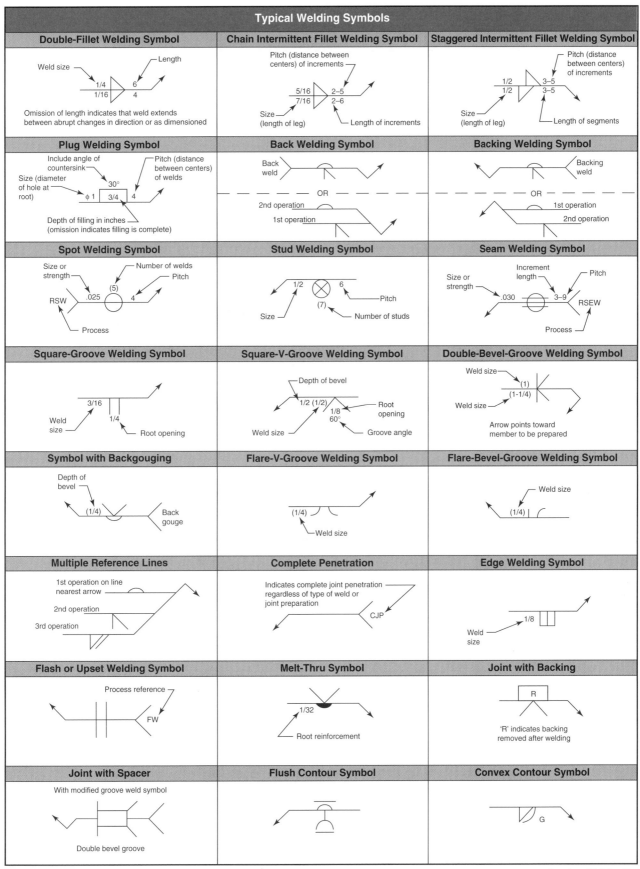

Chart R-4. Typical welding symbols.

Typical Arc Voltages for Short Circuiting Transfer with .035 Inch Diameter Wire			
Metal	**Argon**	**75% Argon/ 25% Carbon Dioxide**	**Carbon Dioxide**
Aluminum	19	—	—
Magnesium	16	—	—
Carbon Steel	—	19	20
Low Alloy Steel	—	19	20
Stainless Steel	—	21	—
Copper	24	—	—
Copper Nickel	23	—	—
Silicon Bronze	23	—	—
Aluminum Bronze	23	—	—
Phosphor Bronze	23	—	—

Goodheart-Willcox Publisher

Chart R-5. Typical arc voltages for short circuiting transfer with .035″ wire.

GMAW Process Parameters				
Steel Material		**.030 in. Diameter Wire**	**Short Circuiting Mode**	
Thickness	**Gas**	**Amperes**	**Wire Speed**	**Volts**
22 ga.	75 Ar–25 CO_2	40–55	90–100	15–16
20 ga.	75 Ar–25 CO_2	50–60	120–135	15–16
18 ga.	75 Ar–25 CO_2	70–80	150–175	16–17
16 ga.	75 Ar–25 CO_2	90–110	220–250	17–18
14 ga.	75 Ar–25 CO_2	120–130	250–340	17–18

Goodheart-Willcox Publisher

Chart R-6a. GMAW process parameters, short circuiting mode.

GMAW Process Parameters				
Steel Material		**.035 in. Diameter Wire**	**Short Circuiting Mode**	
Thickness	**Gas**	**Amperes**	**Wire Speed**	**Volts**
1/8 in.	75 Ar–25 CO_2	140–150	280–300	18–19
3/16 in.	75 Ar–25 CO_2	160–170	320–340	18–19
1/4 in.	75 Ar–25 CO_2	180–190	360–380	21–22
5/16 in.	75 Ar–25 CO_2	200–210	400–420	21–22
3/8 in.	75 Ar–25 CO_2	220–250	420–520	23–24

Goodheart-Willcox Publisher

Chart R-6b.

GMAW Process Parameters				
Steel Material		**.030 in. Diameter Wire**	**Short Circuiting Mode**	
Thickness	**Gas**	**Amperes**	**Wire Speed**	**Volts**
22 ga.	CO_2	40–55	90–100	16–17
20 ga.	CO_2	50–60	120–135	17–18
18 ga.	CO_2	70–80	150–175	18–19
16 ga.	CO_2	90–110	220–250	19–20
14 ga.	CO_2	120–130	250–340	20–21

Goodheart-Willcox Publisher

Chart R-6c.

GMAW Process Parameters				
Steel Material		.035 in. Diameter Wire	Short Circuiting Mode	
Thickness	Gas	Amperes	Wire Speed	Volts
1/8 in.	CO_2	140–150	280–300	21–22
3/16 in.	CO_2	160–170	320–340	21–22
1/4 in.	CO_2	180–190	360–380	23–24
5/16 in.	CO_2	200–210	400–420	23–24

Goodheart-Willcox Publisher

Chart R-6d.

GMAW Process Parameters				
Steel Material		.045 in. Diameter Wire	Short Circuiting Mode	
Thickness	Gas	Amperes	Wire Speed	Volts
1/8 in.	75 Ar–25 CO_2	140–150	140–150	18–19
3/16 in.	75 Ar–25 CO_2	160–170	160–175	18–19
1/4 in.	75 Ar–25 CO_2	180–190	185–195	21–22
5/16 in.	75 Ar–25 CO_2	200–210	210–220	21–22
3/8 in.	75 Ar–25 CO_2	220–250	220–270	23–24

Goodheart-Willcox Publisher

Chart R-6e.

GMAW Process Parameters				
Stainless Steel Material		.035 in. Diameter Wire	Short Circuiting Mode	
Thickness	Gas	Amperes	Wire Speed	Volts
18 ga.	Tri–Mix	50–60	120–150	19–20
16 ga.	Tri–Mix	70–80	180–205	19–20
14 ga.	Tri–Mix	90–110	230–275	20–21
12 ga.	Tri–Mix	120–130	300–325	20–21
3/16 in.	Tri–Mix	140–150	350–375	21–22
1/4 in.	Tri–Mix	160–170	400–425	21–22
5/16 in.	Tri–Mix	180–190	450–475	21–22

Goodheart-Willcox Publisher

Chart R-6f.

GMAW Process Parameters				
Steel Material		.035 in. Diameter Wire	Spray Transfer Mode	
Thickness	Gas	Amperes	Wire Speed	Volts
1/8 in.	98 Ar–2 O_2	160–170	320–340	23–24
3/16 in.	98 Ar–2 O_2	180–190	360–380	24–25
1/4 in.	98 Ar–2 O_2	200–210	400–420	24–25
5/16 in.	98 Ar–2 O_2	220–250	420–520	25–26

Goodheart-Willcox Publisher

Chart R-7a. GMAW process parameters, spray transfer mode.

GMAW Process Parameters				
Steel Material		.045 in. Diameter Wire	Spray Transfer Mode	
Thickness	Gas	Amperes	Wire Speed	Volts
1/8 in.	98 Ar–2 O_2	160–170	160–175	23–24
3/16 in.	98 Ar–2 O_2	180–190	185–195	24–25
1/4 in.	98 Ar–2 O_2	200–210	210–220	24–25
5/16 in.	98 Ar–2 O_2	220–250	220–270	25–26
3/8 in.	98 Ar–2 O_2	300 up	375 up	26–27
1/2 in.	98 Ar–2 O_2	315 up	390 up	29–30

Goodheart-Willcox Publisher

Chart R-7b.

GMAW Process Parameters				
Stainless Steel Material		.035 in. Diameter Wire	Spray Transfer Mode	
Thickness	Gas	Amperes	Wire Speed	Volts
3/16 in.	98 Ar–2 O_2	160–170	400–425	23–24
1/4 in.	Tri–Mix	180–190	450–475	24–25

Goodheart-Willcox Publisher

Chart R-7c.

GMAW Process Parameters				
Stainless Steel Material		1/16 in. Diameter Wire	Spray Transfer Mode	
Thickness	Gas	Amperes	Wire Speed	Volts
3/8 in.	98 Ar–2 O_2	220–250	As Req'd	25–26
7/16 in.	98 Ar–2 O_2	300 up	As Req'd	26–27
1/2 in.	98 Ar–2 O_2	325 up	As Req'd	27–32

Goodheart-Willcox Publisher

Chart R-7d.

GMAW Process Parameters				
Aluminum Material		.035 in. Diameter Wire	Spray Transfer Mode	
Thickness	Gas	Amperes	Wire Speed	Volts
1/8 in.	Argon	110–130	350–400	21–22
3/16 in.	Argon	140–150	425–450	23–24

Goodheart-Willcox Publisher

Chart R-7e.

GMAW Process Parameters				
Aluminum Material		3/64 in. Diameter Wire	Spray Transfer Mode	
Thickness	Gas	Amperes	Wire Speed	Volts
3/16 in.	Argon	140–150	300–325	23–24
1/4 in.	Argon	180–210	350–375	24–25
5/16 in.	Argon	200–230	400–425	26–27
3/8 in.	Argon	220–250	450–480	26–28

Goodheart-Willcox Publisher

Chart R-7f.

GMAW Process Parameters				
Aluminum Material		1/16 in. Diameter Wire	Spray Transfer Mode	
Thickness	Gas	Amperes	Wire Speed	Volts
1/4 in.	Argon	180–210	170–185	24–25
5/16 in.	Argon	200–230	200–210	26–27
3/8 in.	Argon	220–250	220–230	26–28
7/16 in.	Argon	280 up	240–270	28–29
1/2 in.	Argon	300 up	290–300	29–30

Goodheart-Willcox Publisher

Chart R-7g.

GMAW Process Parameters				
Aluminum Material		.030 in. Diameter Wire	Spray Transfer Mode	
Thickness	Gas	Amperes	Wire Speed	Volts
.062 in.	Argon	90	365	18
.125 in.	Argon	125	440	20

Goodheart-Willcox Publisher

Chart R-7h.

Carbon and Low-Alloy Steel Electrodes for FCAW per AWS A5.20					
Electrode	Charpy V-notch, ft./lb (°F)	Shielding	Current	Single/ multipass	Applications
E70T-1	20(0)	CO_2 or 75Ar/25CO_2	DCEP	multi	General-purpose flat and horizontal welding. Railcar fabrication, beams and girders, ships, over-the-road vehicles, storage tanks.
E70T-2	—	CO_2 75Ar/25CO_2	DCEP	single	For single-pass welds on rusted, contaminated base material. Castings, machine bases, oil-field equipment, railcars.
E70T-3	—	self	DCEP	single	Rapid, automated welding on thin-gage steel.
E70T-4	—	self	DCEP	multi	
E70T-5	20 (−20)	CO_2 75Ar/25CO_2	DCEP	multi	Basic slag for good impact properties, low weld metal hydrogen, low crack sensitivity.
E70T-6	20 (−20)	self	DCEP	multi	
E70T-7	—	self	DCEN	multi	
E70T-10	—	self	DCEN	multi	
E70T-G	—	b.	b.	b.	
E71T-1	20(0)	CO_2 75Ar/25CO_2	DCEP	multi	General-purpose all-position welding. Use at low currents (150 A–200 A) to bridge wide gaps, at higher currents (200 A–250 A) for DCEP penetration. Barges, oil rigs, storage vessels, earth-moving equipment.
E71T-5	20 (−20)	CO_2 75Ar/25CO_2	DCEP	multi	
E71T-7	—	self	DCEN	multi	
E71T-8	20 (−20)	self	DCEN	multi	
E71T-11	—	self	DCEN	multi	Thin-gage material, structural steel
E71T-GS	—	a.	b.	b.	
a. Argon additions may improve weld metal properties and welding characteristics. b. New or proprietary wire, properties as specified by the supplier.					

Goodheart-Willcox Publisher

Chart R-8.

Various Electrode Sizes for FCAW		
Thousandths/ inches	Fractional/ inches	Metric/ millimeters
0.030 0.035		0.8 0.9
0.045 0.052	3/64	1.2 1.3
0.062 0.068	1/16	1.6
0.072 0.078	5/64	1.98
0.093 0.109	3/32 7/64	2.4 2.8
0.120 0.128	1/8	3.2
	5/32 3/16	4.0 4.8

Goodheart-Willcox Publisher

Chart R-9. Nominal sizes for FCAW electrodes.

FCAW Steel Electrodes	
Diameter	Inches per Pound*
0.035″ (0.9 mm)	4,350
0.045″ (1.2 mm)	2,500
1/16″ (1.6 mm)	1,300
5/64″ (2.0 mm)	925
3/32″ (2.4 mm)	615
7/64″ (2.8 mm)	550
0.120″ (3.0 mm)	420
*Approximate—Values vary with AWS class and wire type.	

Goodheart-Willcox Publisher

Chart R-10. FCAW steel electrodes.

Typical Operating Parameters for Carbon Steels (without Shielding Gas)		
Wire Diameter	Amperage	Voltage
0.035″ 0.035″ Optimum		
0.045″ 0.045″ Optimum	95–180 130	15–17 15
0.052″ 0.052″ Optimum		
1/16″ 1/16″ Optimum	100–300 200	18–20 19
0.068″ 0.068″ Optimum	120–275 190	17–22 18
0.072″ 0.072″ Optimum	130–250 255	16–25 20–23
5/64″ 5/64″ Optimum	150–300 235	21 20
3/32″ 3/32″ Optimum	245–400 305	24 21
7/64″ 7/64″ Optimum	325–625 450	23–25 27
0.120″ 0.120″ Optimum	400–550 450	28–31 29

Notes:
1. 0.35″ Dia. ES/O 1/2″
2. 0.45″ Dia. ES/O 3/4″
3. 1/16″ Dia. and up ES/O 3/4″
4. 0.120″ Dia. ES/O 2 3/4″
5. Use lower end of range for thinner materials.
6. Use middle of range for vertical welding.
7. Use upper range for flat and horizontal welding only.
8. Use DCRP-DCSP as specified by electrode manufacturer.

Goodheart-Willcox Publisher

Chart R-11. Operating parameters for carbon steels without shielding gas.

Typical Operating Parameters for Carbon Steels (with Shielding Gas)			
Wire Diameter	**Amperage**	**Voltage**	**Shielding**
0.035″ 0.035″ Optimum	130–280 200	20–30 27	CO_2 or 75Ar/25CO_2
0.045″ 0.045″ Optimum	150–290 240	23–30 26	CO_2 or 75Ar/25CO_2
0.052″ 0.052″ Optimum	180–310 260	24–32 26	CO_2 or 75Ar/25CO_2
1/16″ 1/16″ Optimum	180–400 275	25–34 28	CO_2 or 75Ar/25CO_2
5/64″ 5/64″ Optimum	250–400 350	26–30 27	CO_2 or 75Ar/25CO_2
3/32″ 3/32″ Optimum	350–550 450	26–33 31	CO_2 or 75Ar/25CO_2
7/64″ 7/64″ Optimum	500–700 625	30–35 34	CO_2 or 75Ar/25CO_2
1/8″ 1/8″ Optimum	600–800 725	31–36 34	CO_2 or 75Ar/25CO_2

Notes:
1. Use standard CV welding machine.
2. All procedures use DCRP.
3. Electrical stickout:
0.035″–1/16″ diameter electrodes 3/8″ to 3/4″.
5/64″–1/8″ diameter electrodes 3/4″ to 1 1/4″.
4. Voltages may be reduced 1 1/2 V for 75/25 mixture.
5. Use lower end of range for thinner materials.
6. Use lower end of range for uphill welding.
7. Use middle of range for overhead position welding.
8. Use optimum parameters for flat position welding.
9. Use 5/64″–3/32″–7/64″–1/8″ for flat only.

Goodheart-Willcox Publisher

Chart R-12. Operating parameters for carbon steels with shielding gas.

Stainless Steel Flux Cored Electrodes

AWS A5.22 Standard Product Alloy	Variation Product Alloy	Flat Uphill			Out of Position	High Deposition		Applications
		0.035″	0.045″	0.063″	0.045″	0.045″	0.063″	
E308 T-1		•	•	•				Most often used to weld base metal of similar analysis such as AISI 301, 302, 304, 305, and 308.
E308L T-1		•	•	•	•	•	•	Low carbon content improves intergranular corrosion resistance over E308 type. General applications use.
	*308LA T-1		•	•				Primarily used for equipment & piping carrying liquefied helium, nitrogen, and natural gas. Good impact properties at cryogenic temperatures.
	308LN T-1		•					Used to weld nitrogen alloy AISI 304LN.
E309 T-1		•	•	•				Used to weld similar alloys in wrought or cast form and AISI 304 in severe corrosion conditions. Also used to weld AISI 304 to carbon steel.
	309Mo T-1	•	•	•				The addition of Mo to E309 allows use as a first layer cladding on carbon steel which will have further overlays. Also used for joining CrNi and CrNiMo stainless steels to each other and to carbon steel.
E309L T-1		•	•	•	•	•	•	Low carbon content gives better resistance to intergranular corrosion than does E309 T-1.
	309MoL T-1	•	•	•	•	•	•	Typically used to join dissimilar CrNi and CrNiMo stainless steels to each other and to carbon steel and as first layer on carbon steel to be clad with stainless overlay. Low carbon content improves crack resistance.
E309CbL T-1			•					Used to overlay carbon and low alloy steels and to produce a columbium-stabilized first layer.
E310 T-1			•					Most often used to weld AISI 310, 310S, and similar alloys.
E316 T-1		•	•	•				Typically used for welding similar alloys. The presence of molybdenum provides increased creep resistance at elevated temperatures.
E316L T-1		•	•	•	•	•	•	Low carbon increases the resistance to intergranular corrosion but decreases the elevated temperature properties. Commonly used to join AISI 304, 304L, 316, 316L.
	316CuL T-1		•	•				The addition of copper to E316LT-I provides corrosion resistance in higher concentrations of sulfuric acid.
E317L T-1		•	•	•				Used for welding alloys of similar composition and are usually limited to severe corrosion applications involving sulfuric and sulfurous acids and their salts.
	318L T-1		•					For welding AISI 316Cb and 316Ti alloys.
E347 T-1			•	•				This columbium stabilized alloy is generally used to weld AISI 321 and 347 materials.
	*347L T-1	•	•	•				The low carbon content provides better resistance to intergranular corrosion than does E347T-I. Used for final pass overlay on AISI 321 and 347 materials.
E410 T-G			•					Most commonly used to weld alloys of similar analysis. Also used for surfacing of carbon steels to resist corrosion, erosion, and abrasion.
	2209 T-1		•					For welding Avesta Sheffield 2205 Code Plus Two® and similar alloys, to each other, to carbon steels, and to stainless steels.
	NiCr-3 T-1		•					For welding Alloy 600, Alloy 800 and similar materials to each other, to carbon steels, and to stainless steels.
	NiCrMo-3 T-1		•					For welding highly corrosion-resistant materials such as Alloy 625, Alloy 825, Avesta Sheffield 254 SMO® AL6XN®, and similar 6% Mo type alloys. Also suitable for 317LMn and 9% Ni steel.

*Indicates the variation product falls within the AWS specification for the similar standard product.
Avesta flux cored electrodes are available in a wide selection of alloys and in three usage types. Designed for use with 100% CO_2 or a 75% Argon 25% CO_2 shielding gas, they offer excellent economy in all applications. All alloys and sizes are formulated for flat and uphill use with some out-of-position capability. Some alloys and sizes are also available in all position (AP) and high deposition (HD) formats which provide for even greater economies of use.

Avesta Welding Products, Inc.

Chart R-13. Stainless steel flux cored electrodes.

Typical Operating Parameters for Stainless Steels (without Shielding Gas)

Wire Diameter	Amperage	Voltage	Stickout
0.045″	100–180	26–32	3/8″–5/8″
1/16″	150–275	26–32	1/2″–1″
5/64″	200–300	26–32	3/4″–1 1/4″
3/32″	225–350	26–30	1″–1 1/2″
Note: All procedures use DCRP.			

Goodheart-Willcox Publisher

Chart R-14. Operating parameters for stainless steels without shielding gas.

Typical Operating Parameters for Stainless Steels (with Shielding Gas)

Wire Diameter	Amperage	Voltage	Shielding
0.035″	60–150	23–29	CO_2 or 75Ar/25CO_2
0.035″ Optimum	110	26	75Ar/25CO_2
0.045″	85–225	23–32	CO_2 or 75Ar/25CO_2
0.045″ Optimum	160	27	75Ar/25CO_2
1/16″	160–330	25–36	CO_2 or 75Ar/25CO_2
1/16″ Optimum	250	30	75Ar/25CO_2

Notes:
1. Use standard CV welding machine.
2. All procedures use DCRP.
3. Maintain 1/2″ to 3/4″ stickout.
4. Use 1/2″ stickout for overhead position.
5. Ar 75%–CO_2 25% gas mixtures will have lower spatter.
6. Use backhand technique for all positions.
7. Shielding gas flow rates 25–50 cfh.

Goodheart-Willcox Publisher

Chart R-15. Operating parameters for stainless steels with shielding gas.

Preheat Recommendation Chart							
Steel Group	**Steel Designation**		**Carbon**	**Preheat (°F) (a)** **Base Metal 4″ thick**	**Steel Group**	**Steel Designation**		**Carbon**	**Preheat (°F) (a)** **Base Metal 4″ thick**

Steel Group	Steel Designation		Carbon	Preheat (°F) (a) Base Metal 4″ thick	Steel Group	Steel Designation		Carbon	Preheat (°F) (a) Base Metal 4″ thick
Carbon steels	AISI-SAE (c)	1015	0.13–0.18	150°	Chromium steels	AISI-SAE	5015	0.12–0.17	200°
		1020	0.18–0.23	150°			5046	0.43–0.48	450°
		1030	0.28–0.34	200°			5115	0.13–0.18	200°
		1040	0.37–0.44	300°			5145	0.43–0.48	450°
		1080	0.75–0.88	600°			5160	0.56–0.64	550°
Manganese steels	AISI-SAE	1330	0.28–0.33	250°	Austenitic manganese and chrome-nickel steels	ASTM	11%–14% Mn	0.5–1.3	Preheat only to remove chill from base metal.
		1335	0.33–0.38	300°			302	0.15 Max.	
		1340	0.38–0.43	350°			309	0.20 Max.	
		1345	0.43–0.48	400°			310	0.25 Max.	
		1345H	0.42–0.49	400°			347	0.08 Max.	(b)
Molybdenum steels	AISI-SAE	4027H	0.24–0.30	250°	Carbon steel plate structural quality	ASTM	A36	0.27 Max.	250°
		4032H	0.29–0.35	300°			A131 Gr. B	0.21 Max.	200°
		4037H	0.34–0.41	350°			A284 Gr. C	0.29 Max.	250°
		4042H	0.39–0.46	400°			(d) A678 Gr. B	0.20 Max.	200°
		4047H	0.44–0.51	450°	High-strength low-alloy steels structural quality	ASTM	A131-H.S.	0.18 Max.	350°
Chrome molybdenum steels	AISI-SAE	4118	0.17–0.23	250°			A242 Type 2	0.20 Max.	200°
		4130	0.27–0.34	300°			A441	0.22 Max.	200°
		4135	0.32–0.39	400°			A588 Gr. B	0.20 Max.	300°
		4145	0.41–0.49	500°			A633 Gr. E	0.22 Max.	250°
		4145H	0.42–0.49	500°	Alloy and pressure vessel quality steels	ASTM	(d) A514 Gr. F	0.10–0.21	350°
Ni-chrome molybdenum and Ni-moly. steels	AISI-SAE	4340	0.38–0.43	500°			(d) A514 Gr. H	0.12–0.21	300°
		4615	0.18–0.18	250°			(d) A514 Gr. Q	0.14–0.21	550°
		4620	0.17–0.22	250°			A515 Gr. 70	0.35 Max.	300°
		4720H	0.17–0.23	300°			A516 Gr. 70	0.30 Max.	250°
		4820H	0.17–0.23	300°					

Notes:
a. These suggested preheats are recommended when low-hydrogen processes are used on base metals that are 4″ thick. Lower preheats would be needed on thinner material, while higher preheats would be necessary on thicker materials. When using non-low-hydrogen processes, increase suggested preheats by 300°F.
The steels shown on the chart are only partially representative of the steels used in the manufacture of earthmoving and other machinery. A preheat calculator (WC-8) available from Lincoln Electric Co. makes it possible to figure suggested preheats for other steels based upon the chemistry of the steel and the thickness of the parts to be surfaced.
b. It is sometimes advisable to preheat large, thick 11% to 14% manganese parts before welding. Use a maximum of 200°F preheat. (Do not exceed 500°F preheat and/or interpass temperature.) Check base metal with magnet. 11%–14% manganese and the ASTM 300 series of chrome-nickel steels are *not* magnetic.
c. Low-carbon steel.
d. Q & T Steels—see "Need for Preheat."

Lincoln Electric Company

Chart R-16. Preheat recommendation chart.

Preheat, Interpass, and Postheat Temperatures				
AWS Classification[a]	**Preheat and Interpass Temperature[b]**		**PWHT Temperature**	
	°F	°C	°F	°C
E70T5-A1 E80T1-A1 E81T1-A1 E80T5-Ni1 E80T5-Ni2[c] E80T5-Ni3[c] E90T5-N3[c] E90T5-D2 D100T5-D2	300 ± 25	150 ± 15	1150 ± 25	620 ± 15
E81T1-B1 E80T1-B2 E81T1-B2 E80T5-B2 E80T1-B2H E80T5-B2L E90T1-B3 E91T1-B3 E90T5-B3 E100T1-B3 E90T1-B3H E90T1-B3L	350 ± 25	176 ± 15	1275 ± 25	690 ± 15
E71T8-Ni1 E80T1-Ni1 E81T1-Ni1 E71T8-Ni2 E80T1-Ni2 E81T1-Ni2 E90T1-Ni2 E91T1-Ni2 E91T1-D1 E90T1-D3 E80T5-K1 E70T4-K2 E71T8-K2 E80T1-K2 E80T5-K2 E90T1-K2 E91T1-K2 E90T5-K2 E100T1-K3 E100T5-K3 E110T1-K3 E110T5-K3 E111T1-K4 E110T5-K4 E120T5-K4 E120T1-K5 E61T8-K6 E71T8-K6 E80T1-W	300 ± 25	150 ± 15	None	None
EXXXTX-G	Conditions as agreed upon between supplier and purchaser.			

Notes:

a. In this table, the digit (i.e., 0 or 1) before the letter 'T' indicates the primary welding position for which the electrode is designed (usability). Refer to Figure 5-2.

b. These temperatures are specified for testing under this specification and are not to be considered as recommendations for preheat and postweld heat treatment in production welding. The requirements for production welding must be determined by the user.

c. Postweld heat treatment temperatures in excess of 1150°F (620°C) will decrease the impact value.

Goodheart-Willcox Publisher

Chart R-17. Preheat, interpass, and postheat temperatures.

Chemical Compositions of Mild Steel and Low-Alloy Steel Wires							
AWS Class	Carbon	Manganese	Silicon	Sulfur	Phosphorus	Molybdenum	Other
E70S–1	.07–.19	.90–1.40	.30–.50	.035	.025	—	—
E70S–2	.06	.90–1.40	.40–.70	.035	.025	—	.05–.15 Ti. .02–.12 Zr. .05–.15 Al.
E70S–3	.06–.15	.90–1.40	.45–.70	.035	.025	—	—
E70S–4	.07–.15	.90–1.40	.65–.85	.035	.025	—	—
E70S–5	.07–.19	.90–1.40	.30–.60	.035	.025	—	.50–.90 Al.
E70S–6	.07–.15	1.40–1.85	.80–1.15	.035	.025	—	—
E70S–1B	.07–.12	1.60–2.10	.50–.80	.035	.025	.40/.60	—

Goodheart-Willcox Publisher

Chart R-18. Chemical compositions of mild steel and low-alloy steel wires.

Chemical Treatments for Removal of Oxide Films from Aluminum Surfaces			
Solution	Concentration	Procedure	Purpose
Nitric acid	50% water, 50% nitric acid, technical grade.	Immersion 15 min. Rinse in cold water, then in hot water. Dry.	Removal of thin oxide film for fusion welding.
Sodium hydroxide (caustic soda) followed by	5% sodium hydroxide in water.	Immersion 10–60 seconds. Rinse in cold water.	Removal of thick oxide film for all welding processes.
Nitric acid	Concentrated	Immerse for 30 seconds. Rinse in cold water, then hot water. Dry.	Removal of thick oxide film for all welding processes.
Sulfuric-chromic	H_2SO_4 1 gal CrO_3 45 oz Water 9 gal	Dip for 2–3 min. Rinse in cold water, then hot water. Dry.	Removal of films and stains from heat treating, and oxide coatings.
Phosphoric-chromic	H_3PO_3 (75%) 3.5 gal CrO_3 1.75 lb Water 10 gal	Dip for 5–10 min. Rinse in cold water. Rinse in hot water. Dry.	Removal of anodic coatings.

Goodheart-Willcox Publisher

Chart R-19. Chemical treatments for removal of oxide films from aluminum surfaces.

Etching Reagents for Microscopic Examination of Iron and Steel			
Application	**Etching**	**Composition**	**Remarks**
Carbon low-alloy and medium-alloy steels	Nital	Nitric acid (sp gr 1.42) . . . . 1–5 ml Ethyl or methyl alcohol . . 95–99 ml	Darkens perlite and gives contrast between adjacent colonies; reveals ferrite boundaries; differentiates ferrite from martensite; shows case depth of nitrided steel. Etching time: 5–60 sec.
	Picral	Picric acid . . . . 4 g Methyl alcohol . . . . 100 ml	Used for annealed and quench-hardened carbon and alloy steel. Not as effective as No. 1 for revealing ferrite grain boundaries. Etching time: 5–120 sec.
	Hydrochloric and picral acids	Hydrochloric acid . . . . 5 ml Picric acid . . . . 1 g Methyl alcohol . . . . 100 ml	Reveals austenitic grain size in both quenched and quenched-tempered-steels.
Alloy and stainless steel	Mixed acids	Nitric acid . . . . 10 ml Hydrochloric acid . . . . 20 ml Glycerol . . . . 20 ml Hydrogen peroxide . . . . 10 ml	Iron-chromium-nickel-manganese alloy steel. Etching: Use fresh acid.
	Ferric chloride	Ferric chloride . . . . 5 g Hydrochloric acid . . . . 20 ml Water, distilled . . . . 100 ml	Reveals structure of stainless and austenitic nickel steels.
	Marble's reagent	Cupric sulfate . . . . 4 g Hydrochloric acid . . . . 20 ml Water, distilled . . . . 20 ml	Reveals structure of various stainless steels.
High-speed steels	Snyder-Graff	Hydrochloric acid . . . . 9 ml Nitric acid . . . . 9 ml Methyl alcohol . . . . 100 ml	Reveals grain size of quenched and tempered high-speed steels. Etching time: 15 sec. to 5 min.

Goodheart-Willcox Publisher

Chart R-20. Etching reagents for microscopic examination of iron and steel.

Etching Procedures			
Reagents	**Composition**	**Procedure**	**Uses**
Solutions for Aluminum			
Sodium hydroxide	NaOH ...... 1 g Acetic acid ...... 20 ml	Swab 10 seconds.	General microscopic.
Tucker's etch	HF...... 15 ml HCl...... 45 ml HNO_3 ...... 15 ml H_2O ...... 25 ml	Etch by immersion.	Macroscopic.
Solutions for Stainless Steel			
Nitric and acetic acids	HNO_3 ...... 30 ml Acetic acid ...... 20 ml	Apply by swabbing.	Stainless alloys and others high in nickel or cobalt.
Cupric sulfate	$CuSO_4$ ...... 4 gm HCl...... 20 ml H_2O ...... 20 ml	Etch by immersion. 10p11.75	Structure of stainless steels.
Cupric chloride and hydrochloric acid	$CuCl_2$ ...... 5 gm HCl...... 100 ml Ethyl alcohol...... 100 ml H_2O ...... 100 ml	Use cold immersion or swabbing.	Austenitic and ferritic steels.
Solutions for Copper and Brass			
Ammonium hydroxide & ammonium persulfate	NH_4OH...... 1 part H_2O ...... 1 part $(NH_4)_2S_2O_8$ (2 1/2%)...... 2 parts	Immersion.	Polish attack of copper and some alloys.
Chromic acid	Saturated aqueous solution (CrO_3)	Immersion or swabbing.	Copper, brass, bronze, nickel, silver (plain etch).
Ferric chloride	$FeCl_3$ ...... 5 parts HCl ...... 10 parts H_2O ...... 100 parts	Immersion or swabbing (etch lightly).	Copper, brass, bronze, aluminum bronze.
Solutions for Iron and Steel			
Macro Examination			
Nitric acid	HNO_3 ...... 5 ml H_2O ...... 95 ml	Immerse 30 to 60 seconds.	Shows structure of welds.
Ammonium persulfate	$(NH_4)_2S_2O_3$ ...... 10 gms H_2O ...... 90 ml	Surface should be rubbed with cotton during etching.	Brings out grain structure, recrystallization at welds.
Nital	HNO_3 ...... 5 ml Ethyl alcohol...... 95 ml	Etch 5 min. followed by 1 sec. in HCl (10%).	Shows cleanness, depth of hardening, carburized or decarburized surfaces, etc.
Micro Examination			
Picric acid (picral)	Picric acid ...... 4 gms Ethyl or methyl alcohol (95%) .... 100 ml	Etching time a few seconds to a minute or more.	For all grades of carbon steels.

Goodheart-Willcox Publisher

Chart R-21. Etching procedures.

| Brinell | | Vickers or Firth Hardness No. | Rockwell | | Scleroscope | Tensile Strength 1000 psi |
Dia. in mm, 3000 kg Load 10 mm Ball	Hardness No.		C 150 kg Load 120° Diamond Cone	B 100 kg Load 1/16 in. dia. Ball		
2.05	898					440
2.10	857					420
2.15	817					401
2.20	780	1150	70		106	384
2.25	745	1050	68		100	368
2.30	712	960	66		95	352
2.35	682	885	64		91	337
2.40	653	820	62		87	324
2.45	627	765	60		84	311
2.50	601	717	58		81	298
2.55	578	675	57		78	287
2.60	555	633	55	120	75	276
2.65	534	598	53	119	72	266
2.70	514	567	52	119	70	256
2.75	495	540	50	117	67	247
2.80	477	515	49	117	65	238
2.85	461	494	47	116	63	229
2.90	444	472	46	115	61	220
2.95	429	454	45	115	59	212
3.00	415	437	44	114	57	204
3.05	401	420	42	113	55	196
3.10	388	404	41	112	54	189
3.15	375	389	40	112	52	182
3.20	363	375	38	110	51	176
3.25	352	363	37	110	49	170
3.30	341	350	36	109	48	165
3.35	331	339	35	109	46	160
3.40	321	327	34	108	45	155
3.45	311	316	33	108	44	150
3.50	302	305	32	107	43	146
3.55	293	296	31	106	42	142
3.60	285	287	30	105	40	138
3.65	277	279	29	104	39	134
3.70	269	270	28	104	38	131
3.75	262	263	26	103	37	128
3.80	255	256	25	102	37	125
3.85	248	248	24	102	36	122
3.90	241	241	23	100	35	119
3.95	235	235	22	99	34	116
4.00	229	229	21	98	33	113
4.05	223	223	20	97	32	110
4.10	217	217	18	96	31	107
4.15	212	212	17	96	31	104
4.20	207	207	16	95	30	101
4.25	202	202	15	94	30	99
4.30	197	197	13	93	29	97
4.35	192	192	12	92	28	95
4.40	187	187	10	91	28	93
4.45	183	183	9	90	27	91
4.50	179	179	8	89	27	89
4.55	174	174	7	88	26	87
4.60	170	170	6	87	26	85

Goodheart-Willcox Publisher

Chart R-22. Hardness conversion table.

(Continued)

Hardness Conversion Table						
Brinell		**Vickers or Firth Hardness No.**	**Rockwell**		**Scleroscope**	**Tensile Strength 1000 psi**
Dia. in mm, 3000 kg Load 10 mm Ball	**Hardness No.**		**C 150 kg Load 120° Diamond Cone**	**B 100 kg Load 1/16 in. dia. Ball**		
4.65	166	166	4	86	25	83
4.70	163	163	3	85	25	82
4.75	159	159	2	84	24	80
4.80	156	156	1	83	24	78
4.85	153	153		82	23	76
4.90	149	149		81	23	75
4.95	146	146		80	22	74
5.00	143	143		79	22	72
5.05	140	140		78	21	71
5.10	137	137		77	21	70
5.15	134	134		76	21	68
5.20	131	131		74	20	66
5.25	128	128		73	20	65
5.30	126	126		72		64
5.35	124	124		71		63
5.40	121	121		70		62
5.45	118	118		69		61
5.50	116	116		68		60
5.55	114	114		67		59
5.60	112	112		66		58
5.65	109	109		65		56
5.70	107	107		64		56
5.75	105	105		62		54
5.80	103	103		61		53
5.85	101	101		60		52
5.90	99	99		59		51
5.95	97	97		57		50
6.00	95	95		56		49

Goodheart-Willcox Publisher

Chart R-22. *(Continued)*

Properties of Elements and Metal Compositions						
Elements	Symbol	Density (Specific Gravity)	Weight per ft³	Specific Heat	Melting Point	
					°C	°F
Aluminum	Al	2.7	166.7	0.212	658.7	1217.7
Antimony	Sb	6.69	418.3	0.049	630	1166
Armco iron	. . .	7.9	490.0	0.115	1535	2795
Carbon	C	2.34	219.1	0.113	3600	6512
Chromium	Cr	6.92	431.9	0.104	1615	3034
Columbium	Cb	7.06	452.54	. . .	1700	3124
Copper	Cu	8.89	555.6	0.092	1083	1981.4
Gold	Au	19.33	1205.0	0.032	1063	1946
Hydrogen	H	0.070*	0.00533	. . .	−259	−434.2
Iridium	Ir	22.42	1400.0	0.032	2300	4172
Iron	Fe	7.865	490.9	0.115	1530	2786
Lead	Pb	11.37	708.5	0.030	327	621
Manganese	Mn	7.4	463.2	0.111	1260	2300
Mercury	Hg	13.55	848.84	0.033	−38.7	−37.6
Nickel	Ni	8.80	555.6	0.109	1452	2645.6
Nitrogen	N	0.97*	0.063	. . .	−210	−346
Oxygen	O	1.10*	0.0866	. . .	−218	−360
Phosphorus	P	1.83	146.1	0.19	44	111.2
Platinum	Pt	21.45	1336.0	0.032	1755	3191
Potassium	K	0.87	54.3	0.170	62.3	144.1
Silicon	Si	2.49	131.1	0.175	1420	2588
Silver	Ag	10.5	655.5	0.055	960.5	1761
Sodium	Na	0.971	60.6	0.253	97.5	207.5
Sulfur	S	1.95	128.0	0.173	119.2	246
Tin	Sn	7.30	455.7	0.054	231.9	449.5
Titanium	Ti	5.3	218.5	0.010	1795	3263
Tungsten	W	17.5	1186.0	0.034	3000	5432
Uranium	U	18.7	1167.0	0.028		
Vanadium	V	6.0	343.3	0.115	1720	3128
Zinc	Zn	7.19	443.2	0.093	419	786.2
Bronze (90% Cu 10% Sn)	. . .	8.78	548.0	. . .	850–1000	1562–1832
Brass (90% Cu 10% Zn)	. . .	8.60	540.0	. . .	1020–1030	1868–1886
Brass (70% Cu 30% Zn)	. . .	8.44	527.0	. . .	900–940	1652–1724
Cast pig iron	. . .	7.1	443.2	. . .	1100–1250	2012–2282
Open-hearth steel	. . .	7.8	486.9	. . .	1350–1530	2462–2786
Wrought-iron bars	. . .	7.8	486.9	. . .	1530	2786
*Density compared with air.						

Chart R-23. Properties of elements and metal compositions.

Weights and Expansion Properties of Metals				
Metal	Weight per ft³ (lb)	Weight per m³ (kg)	Expansion per °F rise in temperature (0.0001 in.)	Expansion per °C rise in temperature (0.0001 mm)
Aluminum	165	2643	1.360	62.18
Brass	520	8330	1.052	48.10
Bronze	555	8890	0.986	45.08
Copper	555	8890	0.887	40.55
Gold	1200	19222	0.786	35.94
Iron (cast)	460	736 9	0.556	25.42
Lead	710	11373	1.571	71.83
Nickel	550	8810	0.695	31.78
Platinum	1350	21625	0.479	21.90
Silver	655	10492	1.079	49.33
Steel	490	7849	0.689	31.50

Goodheart-Willcox Publisher

Chart R-24. Weights and expansion properties of metals.

General Metric/US Customary Conversions			
Property	To convert from	To	Multiply by
Acceleration (angular)	revolution per minute squared	rad/s^2	$1.745\ 329 \times 10^{-3}$
Acceleration (linear)	in/min^2	m/s^2	$7.055\ 556 \times 10^{-6}$
	ft/min^2	m/s^2	$8.466\ 667 \times 10^{-5}$
	in/min^2	$mm/s^2 mm/s^2$	$7.055\ 556 \times 10^{-3}$
	ft/min^2	m/s^2	$8.466\ 667 \times 10^{-2}$
	ft/s^2		$3.048\ 000 \times 10^{-1}$
Angle, plane	deg	rad	$1.745\ 329 \times 10^{-2}$
	minute	rad	$2.908\ 882 \times 10^{-4}$
	second	rad	$4.848\ 137 \times 10^{-6}$
Area	in^2	m^2	$6.451\ 600 \times 10^{-4}$
	ft^2	m^2	$9.290\ 304 \times 10^{-2}$
	yd^2	m^2	$8.361\ 274 \times 10^{-1}$
	in^2	mm^2	$6.451\ 600 \times 10^{2}$
	ft^2	mm^2	$9.290\ 304 \times 10^{4}$
	acre (US Survey)	m^2	$4.046\ 873 \times 10^{3}$

Goodheart-Willcox Publisher

Chart R-25. General metric/US customary conversions.

(Continued)

General Metric/US Customary Conversions			
Property	**To convert from**	**To**	**Multiply by**
Density	pound mass per cubic inch pound mass per cubic foot	kg/m^3 kg/m^3	2.767 990 × 10^4 1.601 846 × 10
Energy, work, heat, and impact energy	foot pound force foot poundal Btu* calorie* watt hour	J J J J J	1.355 818 4.214 011 × 10^{-2} 1.054 350 × 10^3 4.184 000 3.600 000 × 10^3
Force	kilogram-force pound-force	N N	9.806 650 4.448 222
Impact strength	(see Energy)		
Length	in ft yd rod (US Survey) mile (US Survey)	m m m m km	2.540 000 × 10^{-2} 3.048 000 × 10^{-1} 9.144 000 × 10^{-1} 5.029 210 1.609 347
Mass	pound mass (avdp) metric ton ton (short, 2000 lb/m) slug	kg kg kg kg	4.535 924 × 10^{-1} 1.000 000 × 10^3 9.071 847 × 10^2 1.459 390 × 10
Power	horsepower (550 ft lb/s) horsepower (electric) Btu/min* calorie per minute* foot pound-force per minute	W W W W W	7.456 999 × 10^2 7.460 000 × 10^2 1.757 250 × 10 6.973 333 × 10^{-2} 2.259 697 × 10^{-2}
Pressure	pound force per square inch bar atmosphere kip/in^2	kPa kPa kPa kPa	6.894 757 1.000 000 × 10^2 1.013 250 × 10^2 6.894 757 × 10^3
Temperature	degree Celsius, t°C degree Fahrenheit, t°F degree Rankine, t°R degree Fahrenheit, t°F kelvin, t$_K$	K K	$t_K = t°C + 273.15$ $t_K = (t°F + 459.67)/1.8$ $t_K = t°R/1.8$ $t°C = (t°F - 32)/1.8$ $t°C = t_K - 273.15$
Tensile strength (stress)	ksi	MPa	6.894 757
Torque	inch pound force foot pound force	N·m N·m	1.129 848 × 10^{-1} 1.355 818
Velocity (angular)	revolution per minute degree per minute revolution per minute	rad/s rad/s deg/min	1.047 198 × 10^{-1} 2.908 882 × 10^{-4} 3.600 000 × 10^2
Velocity (linear)	in/min ft/min in/min ft/min mile/hour	m/s m/s mm/s mm/s km/h	4.233 333 × 10^{-4} 5.080 000 × 10^{-3} 4.233 333 × 10^{-1} 5.080 000 1.609 344
Volume	in^3 ft^3 yd^3 in^3 ft^3 in^3 ft^3 gallon	m^3 m^3 m^3 mm^3 mm^3 L L L	1.638 706 × 10^{-5} 2.831 685 × 10^{-2} 7.645 549 × 10^{-1} 1.638 706 × 10^4 2.831 685 × 10^7 1.638 706 × 10^{-2} 2.831 685 × 10 3.785 412

*Thermochemical

Goodheart-Willcox Publisher

Chart R-25. *(Continued)*

Metric Units for Welding		
Property	**Unit**	**Symbol**
Area dimensions	Square millimeter	mm^2
Current density	Ampere per square millimeter	A/mm^2
Deposition rate	Kilogram per hour	kg/h
Electrical resistivity	Ohm meter	$\Omega \cdot m$
Electrode force (upset, squeeze, hold)	Newton	N
Flow rate (gas and liquid)	Liter per minute	L/min
Fracture toughness	Meganewton meter$^{-3/2}$	$MN \cdot m^{-3/2}$
Impact strength	Joule	$J = N \cdot m$
Linear dimensions	Millimeter	mm
Power density	Watt per square meter	W/m^2
Pressure (gas and liquid)	Kilopascal	$kPa = 1000 \ N/m^2$
Tensile strength	Megapascal	$MPa = 1000 \ 000 \ N/m^2$
Thermal conductivity	Watt per meter Kelvin	$W/(m \cdot K)$
Travel speed	Millimeter per second	mm/s
Volume dimensions	Cubic millimeter	mm^3
Electrode feed rate	Millimeter per second	mm/s

Goodheart-Willcox Publisher

Chart R-26. Metric units for welding.

Converting Measurements for Common Welding Properties			
Property	**To convert from**	**To**	**Multiply by**
Area dimensions (mm^2)*	in^2 mm^2	mm^2 in^2	$6.451\ 600 \times 10^2$ $1.550\ 003 \times 10^{-3}$
Current density (A/mm^2)	A/in^2 a/mm^2	A/mm^2 A/in^2	$1.550\ 003 \times 10^{-3}$ $6.451\ 600 \times 10^2$
Deposition rate** (kg/h)	lb/h kg/h	kg/h lb/h	0.045** 2.2**
Electrical resistivity ($\Omega \cdot$m)	$\Omega \cdot$cm $\Omega \cdot$m	$\Omega \cdot$m $\Omega \cdot$cm	$1.000\ 000 \times 10^{-2}$ $1.000\ 000 \times 10^2$
Electrode force (N)	pound-force kilogram-force N	N N ibf	4.448 222 9.806 650 $2.248\ 089 \times 10^{-1}$
Flow rate (L/min)	ft^3/h gallon per hour gallon per minute cm^3/min L/min cm^3/min	L/min L/min L/min L/min ft^3/min ft^3/min	$4.719\ 475 \times 10^{-1}$ $6.309\ 020 \times 10^{-2}$ 3.785 412 $1.000\ 000 \times 10^{-3}$ 2.118 880 $2.118\ 880 \times 10^{-3}$
Fracture toughness (MN$\cdot$m$^{-3/2}$)	ksi$\cdot$in$^{1/2}$ MN$\cdot$m$^{-3/2}$	MN$\cdot$m$^{-3/2}$ ksi$\cdot$in$^{1/2}$	1.098 855 0.910 038
Heat input (J/m)	J/in J/m	J/m J/in	$3.937\ 008 \times 10$ $2.540\ 000 \times 10^{-2}$
Impact energy	foot pound force	J	1.355 818
Linear measurements (mm)	in ft mm mm	mm mm in ft	$2.540\ 000 \times 10$ $3.048\ 000 \times 10^2$ $3.937\ 008 \times 10^{-2}$ $3.280\ 840 \times 10^{-3}$
Power density (W/m^2)	W/in^2 W/m^2	W/m^2 W/in^2	$1.550\ 003 \times 10^3$ $6.451\ 600 \times 10^{-4}$
Pressure (gas and liquid) (kPa)	psi lb/ft^2 N/mm^2 kPa kPa kPa torr (mm Hg at 0°C) micron (μm Hg at 0°C) kPa kPa	Pa Pa Pa psi il/ft^2 N/mm^2 kPa kPa torr micron	$6.894\ 757 \times 10^3$ $4.788\ 026 \times 10$ $1.000\ 000 \times 10^6$ $1.450\ 377 \times 10^{-1}$ $2.088\ 543 \times 10$ $1.000\ 000 \times 10^{-3}$ $1.333\ 22\ \times 10^{-1}$ $1.333\ 22\ \times 10^{-4}$ $7.500\ 64\ \times 10$ $7.500\ 64\ \times 10^3$
Tensile strength (MPa)	psi lb/ft^2 N/mm^2 MPa MPa MPa	kPa kPa MPa psi lb/ft^2 N/mm^2	6.894 757 $4.788\ 026 \times 10^{-2}$ 1.000 000 $1.450\ 377 \times 10^2$ $2.088\ 543 \times 10^4$ 1.000 000
Thermal conductivity (W[m$\cdot$K])	cal/(cmÃs$\cdot$°c)	W/(m$\cdot$K)	$4.184\ 000 \times 10^2$
Travel speed, electrode feed speed (mm/s)	in/min mm/s	mm/s in/min	$4.233\ 333 \times 10^{-2}$ 2.362 205

*Preferred units are given in parentheses.
**Approximate conversion.

Goodheart-Willcox Publisher

Chart R-27. Converting measurements for common welding properties.

Glossary

A

American Society of Mechanical Engineers (ASME). An organization that has developed eleven code sections pertaining to the design, manufacturing, inspection, and installation of boilers, pressure vessels, nuclear power plants and a separate code for pressure piping applications. (18)

annealing. Softening of metal by heating and slow cooling. (5)

ANSI Z49.1. A national standard, published by the American Welding Society, that outlines welding safety practices for the protection of personnel and property in a detailed manner. (2)

arc blow. The deflection of the intended arc pattern due to magnetism. (8)

arc voltage. The voltage output of the power source when an arc is present. (3)

argon. The inert shielding gas that is most commonly used for GMAW. It is heavier than air and forms a blanket around the electrode and molten metal, reducing spatter and contamination. (4)

arrow side. The area below the AWS welding symbol horizontal reference line. A weld symbol placed there indicates that the weld is to be made on the side of the weld joint touched by the arrow. (6)

austenite. A metallurgical term identifying a specific type of grain structure. (9)

austenitic stainless steels. Stainless steels with 16%–26% chromium content and 8%–24% nickel content. (9)

automatic welding. Welding method in which the welder operator uses specialized mechanical or computer-controlled equipment or robotics to complete the required weld. (1)

AWS welding symbol. A symbol developed by the American Welding Society that is found on drawings to indicate the type of joint, placement, and the type of weld to be made. (6)

B

backhand welding. GMAW operation in which the welding gun points back at the weld and the weld progresses in the opposite direction. Also called *pull welding*. (7, 12)

backing bar. Controls the penetration of the weld bead on the backside of the joint. Since the weld usually penetrates into the backing bar, the bar is not removed after welding. (6)

buildup welds. Welds made to restore worn parts to original dimensions or to apply the surfacing material. (16)

burnback. The problem of an electrode melting and adhering to the contact tip. (10)

burnback timer. A timer placed into the welding circuit to allow the electrical current to flow to the electrode wire after the electrode wire stops feeding. (11)

burn-through. A weld that has melted through the weld joint, resulting in a hole and/or excessive penetration. (7)

buttering. Depositing a material on one or both of the joint pieces to make the joint materials metallurgically compatible with a common filler material. (6) Surfacing an area where materials are to be joined and a dissimilar material is used for the joining. (16)

C

carbide precipitation. A condition in which chromium leaves the grains and combines with carbon in the grain boundary. (9, 14)

carbon dioxide. A compound reactive gas made up of carbon monoxide and oxygen. (4)

carbon steels. A group of steels that contain carbon, manganese, and silicon, along with small amounts of other elements. (8, 13)

cast. The diameter of one complete circle of wire from the spool as it lies on a flat surface. (5)

cast irons. Iron-carbon alloys that contain 93% – 95% iron, 2.5% – 3.5% carbon, and smaller quantities of other elements. (15)

Certified Welding Inspector (CWI). A welding inspector who has industry experience and has completed training and examinations provided by the American Welding Society. (19)

chrome-moly steels. Chromium-molybdenum steels, which are used in applications requiring high strength. (8)

cladding. Overlaying a weld by placing a layer of weld metal on the surface. Also called *surfacing*. (6) The application of a material that provides a corrosion-resistant surface. (16)

code. A national standard that holds legal standing. (18)

constant current (CC). Type of power source in which a small current change occurs from a large arc voltage change. (1)

constant voltage (CV). A type of power source that yields a large welding current change from a small arc voltage change. (1)

contact tips. Part of a welding gun that conducts electrical current from the power source to the consumable welding wire. (3)

D

defects. Extensive weld flaws that may cause a weld to fail. (17)

demurrage. Rental fee charged by the distributor on each gas cylinder used. (4)

deoxidizer. An element that helps remove oxygen and nitrogen from the weld, reducing the occurrence of weld metal porosity. (8)

destructive testing. Series of tests by destruction to determine the physical properties of a weld. (17)

Dewar flasks. Specially constructed tanks, similar to vacuum bottles, used for the storage of liquefied gases. (2)

discontinuities. Weld flaws that are not necessarily defects. (17)

double weld. A weld that is made from both sides of the joint. (6)

drag angles. Angles for backhand welding on linear welds. (12)

ductile iron. Cast iron in which various elements are added to create spheres of the excess carbon during the initial melting of the ore, resulting in an extremely strong and ductile material. Also called *nodular iron* or *spheroidal iron*. (15)

ductility. Ability to deform without breaking. (15)

duplex stainless steels. Stainless steels that are part ferritic, part austenitic. They solidify as 100% ferrite, but half of the ferrite transforms into austenite during cooling. (9)

duty cycle. The percentage of time in a ten-minute period that a machine can run at its rated output without overheating. (3)

E

electrode extension. The distance from the contact tip to the end of the electrode. (7)

*The number in parentheses following each definition indicates the chapter in which the term can be found.

electrode wire. In GMAW, the wire that is melted as a result of the electric arc and becomes part of the weld. (1)

essential variables. Changes that alter the mechanical properties of the completed weldment. (18)

F

ferritic stainless steels. Stainless steels that contain 10.5%–30% chromium and up to 1.20% carbon, along with small amounts of manganese and nickel. (9)

ferromagnetic. Term that describes an iron-based metal with magnetic properties. (17)

fillet weld. A weld that is approximately a triangular cross section used to connect perpendicular or uneven joint surfaces at right angles to each other in a lap joint, T-joint, or corner joint. (6)

fillet weld break test. A destructive test in which a sample fillet weld that is welded on one side only is broken. The sample has a load applied to its nonwelded side, transverse to the weld, and directed to its nonwelded side. The load is increased until the weld fails. (18)

flowmeter. A device that regulates the volume of gas flow through the welding system to the weld zone. (4)

flux cored arc welding (FCAW). Electric arc welding process used to fuse metallic parts by heating them with an arc between a continuously fed welding wire electrode and the workpiece. The electrode is a tubular flux-filled wire. (1)

forehand welding. GMAW operation in which the welding gun points forward in the direction of travel. Also called *push welding*. (7, 12)

free machining steels. Steels that have been modified for use as machining stock by adding special materials to permit the machining operation. (8)

G

gas metal arc welding (GMAW). Electric arc welding process used to fuse metallic parts by heating them with an arc between a continuously fed welding wire electrode and the workpiece. The small-diameter electrode is consumed during the process. (1)

globular transfer. A generally undesirable deposition mode in which the electrode burns off above or in contact with the workpiece in an erratic globular pattern. (1)

gray iron. Commonly used, inexpensive form of cast iron that has a gray appearance and very low ductility. (15)

groove weld. A weld used to connect two joint surfaces or edges against each other. (6)

guided bend test. A destructive test in which a specimen is bent to a specified bend radius to evaluate the ductility and soundness of welded joints. (18)

H

hardfacing. A special form of surfacing applied to make a part harder as a means of reducing wear. (16)

helium. An inert shielding gas that is commonly used as part of the gas mixture for stainless steel and aluminum applications. (4)

helix. The maximum height of any point of a complete circle of wire above the flat surface. (5)

high-carbon steel. Carbon steel that contains 0.30% – 0.59% carbon. (8)

hot cracking. The formation of shrinkage cracks during the solidification of weld metal. (9)

hydrogen. A reactive gas used in small percentages when added to argon for shielding stainless steel and nickel alloys. Its high thermal conductivity helps produce a fluid weld pool and improved wetting action. (4)

I

inductance. The separation of the molten drops of metal from the electrode controlled by squeezing forces exerted on the electrode due to the current flowing through it. (3)

inert gas. A chemically inactive gas, such as argon or helium, that does not react chemically with the molten weld pool. (4)

interference fit. The inside diameter of an outer cylindrical part is made several thousandths of an inch smaller than the outside diameter of the inner part. Before assembly, the outer part is heated until it expands enough to slide over the inner part. As it cools, it locks onto the smaller part. (6)

intermittent (stitch) welds. Short weld beads of one to two inches. (11)

interpass temperature. The minimum/maximum temperature of the weld metal before the next pass in a multipass weld is made. (6, 8, 13)

J

joint. The manner in which materials fit together. (6)

L

layer wound. Electrode wire used for GMAW hard wires. (5)

level-layer wound. Electrode wire used for all soft and cored wire. The electrode is carefully spooled onto the reel with each circle next to the previous one and each layer directly over the one beneath. (5)

liquid penetrant inspection. A nondestructive test performed in which colored liquid dye and fluorescent penetrant are applied to the weld to check for surface flaws. (17)

low-alloy steels. Steels that contain varying amounts of carbon in addition to many alloying elements, including chromium, molybdenum, nickel, vanadium, and manganese. (8)

low-carbon steel. Carbon steel that contains up to 0.14% carbon. (8, 13)

M

macroetch test. Small samples of a welded joint are removed and polished across their cross section and then etched using mild acid mixture. The acid etch provides a clear visual appearance of the internal structure of the weld. (18)

magnetic particle inspection. Nondestructive inspection test using a magnetized part and finely divided iron particles to outline discontinuities in the base metal or the weld. (17)

malleable iron. Cast iron made by heat-treating white iron to change the graphite flakes to a spheroidal shape. This change reduces the tensile strength of the material but improves the ductility. (15)

manifold. A system that allows two or more cylinders to be regulated and connected through a distribution pipeline system to deliver gas to several welding stations as needed. (4)

martensitic stainless steels. Stainless steels that contain 11%–18% chromium and up to 0.20% carbon. (9)

medium-carbon steel. Carbon steel that contains 0.15%–0.29% carbon. (8)

metal cored electrode wire. A tubular electrode that consists of a metal sheath and a core of various powdered materials, primarily iron. The core contributes almost entirely to the deposited weld metal. (8)

N

nitrogen. An inexpensive inert gas used for purging applications only. (4)

nondestructive testing. Inspection of a weld or assembly to verify quality. The part is usable after the testing is done. (17)

nonessential variables. Variables in the welding procedure that have no significant impact on the mechanical properties of the completed weld. (18)

nonferrous metal. A metal that contains no iron. (10)

nozzles. Tubes that attach to the end of the welding gun and direct the shielding gas flow over the weld area. (3)

O

open circuit voltage. The voltage output of the power source before an arc is established. (3)

other side. The area above the AWS welding symbol horizontal reference line. A weld symbol placed there indicates the weld is to be made on the side of the weld joint opposite the side touched by the arrow. (6)

overlap. Weld metal that has flowed over the edge of the joint and improperly fused with the base metal. (17)

oxide. A material formed through the combination of metal and oxygen. (10)

oxygen. A reactive gas that is not used as a single gas for GMAW welding but as part of a gas mixture to attain specific arc patterns. (4)

P

personal protective equipment (PPE). Equipment and protective clothing worn to minimize exposure to a variety of hazards, including sparks, heat, and the arc created during welding. (2)

phosgene gas. Poisonous, colorless, very volatile, suffocating gas that is created when trichloroethylene vapor enters the arc and is heated. (2)

plug weld. A weld made in a circular hole in one member of a joint, fusing that member to another member. (6)

postheating. A controlled cooling rate of the weld. The weldment is kept at a specific temperature for a specific period of time after welding is completed. (8, 13)

power source. The machine that supplies the current and voltage suitable for the welding process. (1)

precipitation hardening stainless steels. Stainless steels that provide an optimum combination of the properties of martensitic and austenitic grades. Properties include high strength and high corrosion resistance achieved by an age hardening heat treatment technique. (9)

preheating. Bringing the base metal up to temperature before the welding process begins. Heat is applied to the weld area to obtain a specific temperature prior to the start of welding. (8)

prequalified joint. A joint design that does not require qualification tests for mechanical values because base material and filler material combinations are specified in the code. (18)

procedure qualification record (PQR). A document containing the test results for qualification of a welding procedure specification. (18)

pull welding. GMAW operation in which the welding gun points back at the weld and the weld progresses in the opposite direction. Also called *backhand welding*. (7)

pulsed spray transfer. A deposition mode in which a drop of molten filler metal is melted off or pulsed from the electrode wire at a controlled rate and time in the weld cycle. (1)

purging. Using a gas to displace the atmosphere from the adjacent weld zone under the weld opening or inside a pipe or tube. Argon, nitrogen, and helium are gases used for this purpose. (4)

push welding. GMAW operation in which the welding gun points forward in the direction of travel. Also called *forehand welding*. (7)

push-pull type feeder. Wire feeder type in which one set of feed drive rollers pushes the wire from the wire feeder to another set of drive rollers mounted in the welding gun. (3)

push-type feeder. Wire feeder type in which one or two sets of feed drive rollers push the wire from the spool to the welding gun. (3)

Q

qualified procedure. A welding sequence that has met all testing requirements included in the fabrication specification. (17)

R

radiographic inspection. A nondestructive test that shows the presence and nature of discontinuities inside a weld. X-rays and gamma rays are used to penetrate the weld and flaws are revealed on exposed film. (17)

reactive gas. A gas, such as oxygen, hydrogen, or carbon dioxide, that combines chemically with the weld pool to produce a desirable effect. (4)

reference line. The horizontal line of the AWS welding symbol. All information about the weld to be made is positioned above, below, or on this line. (6)

regulator. A device that reduces the pressure from the cylinder to the desired manifold pressure level, usually ranging from 20 psi to 50 psi. (4)

regulator/flowmeter. A device that reduces the pressure of the shielding gas and also maintains the volume of gas to the weld zone. (4)

résumé. A brief outline of your education, work experience, and other qualifications for work. (19)

run-off tab. A short piece of the base metal tack welded to the end of the weld joint to allow the weld to continue and end beyond the weldment. The tab is removed after the weld is completed. (10, 14)

run-on tab. An added piece of base material where the weld will be started off the intended weld joint. The tab will be removed after the weld is completed. (14)

S

safety data sheets (SDS). Documents that detail the properties and hazards of chemical products, including precautions such as PPE to be worn and the proper handling, storage, and disposal of substances. (2)

semiautomatic welding. Welding method in which the weld is produced and controlled by a welder using a handheld gun and the electrode wire is fed from a wire feeder system. (1)

sheet metal gauge. A tool used to check the thickness of sheet metal. (11)

shielding gas. A pure gas or a mixture of several gases that perform several functions in the GMAW process, including shielding the electrode and molten metal from contamination from the atmosphere. (1, 4)

short circuiting transfer. Metal deposition mode in which the electrode wire short-circuits when it is fed into and contacts the workpiece, causing the electrode to melt off and be deposited as molten metal into the weld joint. (1)

side bend test. The coupon weld is centered in the fixture with the side of the weld cross section facing upward. The ram of the test fixture is closed until the legs of the coupon are even. The coupons are evaluated after bending for cracks on the outer surface of the bent coupon. (18)

slot weld. A weld made in an elongated hole in one member of a joint, fusing that member to another member. The hole may be open at one end. (6)

soft skills. Interpersonal skills, such as communication skills and ability to be a team member, and self-management skills, such as time-management skills. (19)

spool gun. A gun that includes a small spool of filler metal in the body, an electrical motor, and drive rolls to feed the wire. (3)

spray transfer. A deposition mode in which the electrode is melted off above the workpiece, forming a fine spray of molten metal that is deposited into the weld zone. (1)

stainless steels. Iron-based alloys that contain a minimum of 10.5% chromium. (9, 14)

stepover distance. The distance moved over to make the next weld when, after the first pass is made, the remaining passes are located on half of the completed weld and half of the base metal. (16)

stickout. The length of unmelted electrode extending past the end of the gas nozzle. (7)

stress relief. A form of weldment heating in which the entire part is placed in a furnace for a period of time at a specific temperature. (8)

stringer bead pattern. Pattern of depositing metal in which travel is along the joint with very little side-to-side motion. A small zigzag motion or small circular motion is used. (7, 12)

stringer beads. Welds made without side-to-side motion. (6)

surfacing. Overlaying a weld by placing a layer of weld metal on the surface. Also called *cladding*. (6)

T

temper designations. The basic aluminum heat treatment temper designations are F (fabricated), O (annealed, or the lowest strength), H (strain hardened), and T (solution heat-treated). Additional numbers are added to the designation to describe the treatment process. (10)

transverse tension test. A destructive test in which tensile strength of the welded joint is obtained by pulling specimens to failure. Tensile strength is determined by dividing the maximum load required during testing by the cross-sectional area. (18)

travel angle. Longitudinal gun angle (along the weld). (7)

U

ultrasonic inspection. A nondestructive test method that uses high-frequency sounds to determine interior quality of a weld. (17)

undercut. A groove melted into the base metal next to the toe of the weld and left unfilled by weld metal. (6) A depression that forms at the end of the bead if a dwell (wait period) is not allowed at the end of the pass. (12)

V

visible stickout. The distance between the end of the gas nozzle and the end of the electrode that can be viewed. (12)

visual inspection. Viewing the weld on the front or top surfaces (and penetration side, if visible) plus using measuring tools, scales, squares, mirrors, flashlight and other weld measurement instruments to determine the condition of the weld. (17)

W

weave bead pattern. Pattern of depositing metal in which the weld is wider than a stringer bead and requires a dwell (wait) of the gun at the end of each weave to fill the metal into the weld without undercut. Also called *wash bead pattern* or *oscillation bead pattern*. (7, 12)

weld schedule. A list of parameters and variables for the welding of component parts, including materials, wire sizes, types of gases, and techniques used to make the weld. (7)

weld symbols. Standardized representations used for types and positions of welds, weld finishes, and welding techniques. They are placed in specified areas of the AWS welding symbol. (6)

welder qualification. Qualification based on the welder's ability to perform a proven welding procedure using a specified process and the proper techniques to complete a satisfactory weld. (18)

welder qualification test record. A document containing all of the information from the applicable WPS and PQR used to qualify a welder to that procedure, including the results of the mechanical tests. (18)

welding gun. Device that directs the electrode wire and shielding gas to the weld and contains the trigger switch that controls the wire feeder operations. (1)

welding procedure specification (WPS). A document that provides the required welding variable information for a specific weld application to ensure that the weld can be duplicated by properly trained welders. (18)

welding voltage. Welding parameter that determines the actual arc gap established between the end of the electrode and the workpiece. (7)

whisker. A piece of electrode wire that extends through the completed weld joint. (11)

white iron. Unweldable cast iron with a hard, abrasive surface and very low ductility. (15)

wire feeder. Device that controls the feed speed rate of the electrode wire, the shielding gas solenoid, and the electrical current from the work to the power source. (1)

wire speed. The rate at which the welding wire is fed through the welding gun. (7)

work angle. Transverse (across the weld) gun angle. (7)

Index

A

B

C

D

E

F